T0358303

Fitting Local Volatility

Analytic and Numerical Approaches
in Black-Scholes and
Local Variance Gamma Models

Fitting Local Volatility

Volatility

Analytic and Numerical Approaches
in Black-Scholes and
Local Variance Gamma Models

Andrey Itkin

New York University, USA

 **World Scientific**

NEW JERSEY · LONDON · SINGAPORE · BEIJING · SHANGHAI · HONG KONG · TAIPEI · CHENNAI · TOKYO

Published by

World Scientific Publishing Co. Pte. Ltd.

5 Toh Tuck Link, Singapore 596224

USA office: 27 Warren Street, Suite 401-402, Hackensack, NJ 07601

UK office: 57 Shelton Street, Covent Garden, London WC2H 9HE

Library of Congress Control Number: 2019053436

British Library Cataloguing-in-Publication Data
A catalogue record for this book is available from the British Library.

FITTING LOCAL VOLATILITY
Analytic and Numerical Approaches in Black-Scholes and Local Variance Gamma Models

Copyright © 2020 by World Scientific Publishing Co. Pte. Ltd.

All rights reserved. This book, or parts thereof, may not be reproduced in any form or by any means, electronic or mechanical, including photocopying, recording or any information storage and retrieval system now known or to be invented, without written permission from the publisher.

For photocopying of material in this volume, please pay a copying fee through the Copyright Clearance Center, Inc., 222 Rosewood Drive, Danvers, MA 01923, USA. In this case permission to photocopy is not required from the publisher.

ISBN 978-981-121-276-5

For any available supplementary material, please visit
https://www.worldscientific.com/worldscibooks/10.1142/11623#t=suppl

Desk Editors: Herbert Moses/Shreya Gopi

Typeset by Stallion Press
Email: enquiries@stallionpress.com

Printed in Singapore

Foreword

– By Alex Lipton

Co-Founder and CTO, Silamoney, Partner,
Numeraire Financial, Connection Science Fellow,
Massachusetts Institute of Technology, Cambridge, USA

Prof. Andrey Itkin, a well-known and highly respected financial engineer, gives a *tour de force* performance in this short, insightful book. To understand his contribution properly, we need to step back in history.

Derivatives, including options, forwards, and futures, have been around at least since early modern times. While forwards and futures are linear instruments, which are relatively simple to handle, options are nonlinear in nature and are more difficult to deal with. They come in all kind of flavors and include vanilla calls and puts with hockey-stick payoffs, European options with more general payoffs, American, Bermudan and Asian options, as well as barrier and other exotic options. The corresponding underliers include stocks, bonds, currencies, commodities, etc.

The original scientific approaches to options valuation were developed by Bachelier (1900), Bronzin (1908), Boness (1964), Samuelson (1965), and others. Yet, explosive development of the field is due to the seminal work of Black–Scholes–Merton (BSM), Black and Scholes (1973), and Merton (1973), who developed a novel theory of rational option pricing, which gained universal acceptance and eventually became an indispensable tool of modern financial engineering. The main advantage of their theory compared to its predecessors is its ability to tie together pricing and hedging. By necessity, BSM theory is based on several idealized assumptions, which do not hold in the real world. These assumptions are as follows: (A) markets are efficient and frictionless, i.e., there are no transaction costs and taxes; (B) there are no restrictions on short sales, frequency of trading, or the amount of shares bought and sold; (C) the stock prices are governed by log-normal stochastic processes with time-independent parameters. Provided that these assumptions hold, one can construct a perfect hedge for a contingent claim and price it under a risk-neutral measure. In the BSM framework every stock is characterized by a single number σ, which is called its volatility, while its physical drift μ is unimportant, and can be risk neutralized, so that $\mu \to r - q$, where r is the risk-free interest rate, and q is the dividend rate. Thus, in order to price an option, we can assume that the corresponding stock price is governed by the following stochastic differential equation (SDE):

$$\frac{dS(t)}{S(t)} = (r - q) \, dt + \sigma dW(t), \tag{1}$$

$$S(0) = S_0,$$

where W is a standard Wiener process.

For a European option, the corresponding BSM partial differential equation (PDE) and the terminal condition have the form

$$V_t(t, S) + \frac{1}{2}\sigma^2 S^2 V_{SS}(t, S) + (r - q) S V_S(t, S) - rV(t, S) = 0, \tag{2}$$

$$V(T, S) = v(S),$$

where T is the option maturity, and $v(S)$ is its payoff. For calls and puts, we have

$$v(S) = (\omega(S - K))_+, \tag{3}$$

where

$$(x)_+ = \begin{cases} x, & x \geq 0, \\ 0 & x < 0, \end{cases} \tag{4}$$

and ω is the call/put indicator, $\omega = 1$ for calls, and $\omega = -1$ for puts. It is easy to show that BSM prices of vanilla calls and puts with strike K and maturity T have the form

$$V(t, S, T, K; r, d, \sigma) = \omega e^{-r\tau} \left[F(t, T) \mathfrak{N} \left(\omega \frac{\ln(F(t, T)/K) + \frac{1}{2}\sigma^2\tau}{\sigma\sqrt{\tau}} \right) \right. \tag{5}$$

$$\left. - K\mathfrak{N} \left(\omega \frac{\ln(F(t, T)/K) - \frac{1}{2}\sigma^2\tau}{\sigma\sqrt{\tau}} \right) \right],$$

where $\mathfrak{N}(.)$ is the cumulative normal distribution, $\tau = T - t$, and $F(t, T) = \exp((r - q)\tau) S(t)$ is the forward price. Prices of more complicated options can be computed either analytically or numerically.

However, in real markets, idealized BSM assumptions do not hold. Market prices are such that different vanilla options on a given underlier must be priced using different implied volatilities $\sigma_I(t, S, T, K)$ in Eq. (3), so that every underlier has its implied volatility surface, rather than a single volatility. For a fixed maturity T, this fact results in the volatility skew or smile; for a fixed strike K it manifests itself as the term structure of volatility. As a result, considerable modeling insight is required for pricing vanilla and, especially, exotic options consistently with the market and with each other.

Local volatility models assume that the stock price is governed by Eq. (1), with the constant σ replaced by the so-called local volatility $\sigma_L(t, S)$. These models extend the standard BSM framework in a relatively mild fashion, so that they are complete and allow for a perfect hedge. In general, local volatility models cannot generate proper implied volatility dynamics and tend to misprice exotic options, while at the same time pricing vanillas perfectly. The variance gamma model of Madan and Seneta (1990) can be generalized to the local variance gamma model of Carr and Nadtochiy (2017).

Jump diffusion models, first proposed by Merton (1976), augment the regular diffusion with Poissonian jumps. These models are necessarily incomplete and don't allow for a perfect hedge, since small and large movements of the underlier cannot be handled simultaneously. For many underliers, local volatility models augmented with jumps tend to agree with market prices well.

Stochastic volatility models are useful in many situations. The most popular model of this kind is due to Heston (1993). The model assumes that volatility is driven by a mean-reverting square-root process and generates closed-form expressions for call and put prices. For certain underliers, for example, liquid currency pairs, stochastic volatility models are adequate, though never perfect.

Finally, universal volatility models proposed by Lipton (2002) combine the best features of local, jump diffusion and stochastic volatility models; as a result, they produce reliable prices and hedges not only for calls and puts but also for many exotic options.

A large portion of Itkin's book is devoted to the calibration of local volatility models. The instrument of choice for calibrating these models to the market is the celebrated forward equation of option pricing due to Dupire (1994). An alternative but equivalent formulation is given by Derman and Kani (1994). The Dupire equation allows one to price calls and puts with different strikes and maturities at once. Depending on the context, it can he used to express $\sigma_I(t, S)$ in terms of $\sigma_L(t, S)$, or, conversely, to express $\sigma_L(t, S)$ in terms of $\sigma_I(t, S)$. Specifically, written in terms of call prices $C(t, S, T, K)$, the Dupire equation reads

$$C_T(T, K) - \frac{1}{2}\sigma_L^2(T, K) K^2 C_{KK}(T, K)$$
$$+ (r - q) K C_K(T, K) + qC(T, K) = 0,$$
$$C(0, K) = (S - K)_+. \tag{6}$$

For a given σ_L, Eq. (6) can be solved forward to generate the corresponding $C(T, K)$, and then σ_I via Eq. (5). Alternatively, and more importantly, for a given price surface $C(T, K)$, Eq. (6) can be used to calibrate the corresponding local volatility model and get

$$\sigma_L^2(T, K) = \frac{C_T(T, K) + (r - q) K C_K(T, K) + qC(T, K)}{\frac{1}{2}K^2 C_{KK}(T, K)}. \tag{7}$$

Once σ_L is known, one can use an appropriately modified BSM pricing problem to find prices of more complicated options.

Equation (7) is beautiful in theory, but rather difficult to use in practice for a variety of reasons, first and foremost due to the fact that in real markets $C(T, K)$ is known for a rather sparse set of discrete pairs $\Xi = \{(T_i, K_i)\}$, $i = 1, ..., N$. This book is dedicated to developing novel and technically advanced methods for solving the calibration problem in earnest.

In Part I, the author discusses the general concept of local volatility and introduces the all-important no-arbitrage conditions. He develops various forms of no-arbitrage interpolation which are used in the rest of the book. The advantage of these interpolations is that, in addition to preserving no-arbitrage conditions by construction, they allow closed-form solutions for various special cases of the general volatility problem. In addition, the author discusses a recent work of Carr and Pelts (2015) on no arbitrage for the implied volatility surface, and puts it into the general framework.

In Part II, the calibration problem for a discrete set of maturities and strikes is analyzed in great detail. The author extends the work of Andreasen and Huge (2011), and, more closely, Lipton and Sepp (2011), to produce a continuous piecewise linear local volatility surface $\sigma_L(t, S)$ in the spirit of Itkin and Lipton (2018). He also uses modern regression methods to perform parameter calibration, starting with the Stochastic Volatility Inspired (SVI) local volatility parameterization, Gatheral (2006), and proceeding to his own parameterization, Itkin (2015), which he developed a decade ago while working as a market maker. The author convincingly demonstrates the advantages of this parameterization by comparing the fit with representative market data as well as with SVI.

Finally, in Part III, the author expands the local variance gamma model of Carr and Nadtochiy (2017) in the spirit of Carr and Itkin (2018a, 2018b). He adds a drift and makes the stochastic driver geometric rather than arithmetic by using a clever time change. After that, the author shows how to calibrate the model to the market in an efficient manner by exploiting various new no-arbitrage interpolations introduced in Part I.

This book expands the previous work done by the author, some solo and some in co-authorship with Peter Carr and Alexander Lipton, by providing additional material, which makes the exposition clear and uniform. One of the key advantages of the book is its coverage of the most recent findings in the general area of volatility calibration. The other one is that it is a treasure trove of numerical recipes making the volatility surface calibration both fast and accurate.

Itkin's book can be used as a foundation for an advanced master or PhD course in financial engineering and mathematical finance. It can also be used by practitioners and academics who want to learn the most modern and efficient approaches to building local and implied volatility surfaces in a fast and accurate way. I recommend it wholeheartedly.

References

[1] Andreasen, J. and Huge, B. (2011). Volatility interpolation, *Risk Magazine*, pp. 76–79.
[2] Bachelier, L. (1900). Theorie de la speculation, *Annales de l'Ecole Normale Superieure*, 17, pp. 21–86.
[3] Black, F. and Scholes, M. (1973). The pricing of options and corporate liabilities, *Journal of Political Economy*, 81, pp. 637–659.
[4] Boness, A.J. (1964). Elements of a theory of a stock option value, *Journal of Political Economy*, 72, pp. 163–675.
[5] Bronzin, V. (1908). Theorie der Prämiengeschäfte; Franz Deuticke.
[6] Carr, P. and Itkin, A. (2018a). An expanded local variance gamma model, Available at https://arxiv.org/pdf/1802.09611.pdf.
[7] Carr, P. and Itkin, A. (2018b). Geometric local variance gamma model, Available at https://arxiv.org/abs/1809.07727.
[8] Carr, P. and Nadtochiy, S. (2017). Local Variance Gamma and explicit calibration to option prices, *Mathematical Finance*, 27(1), pp. 151–193.
[9] Carr, P. and Pelts, G. (2015). Duality, deltas, and derivatives pricing, in Steve Shreve's 65th Birthday Conference, http://www.math.cmu.edu/CCF/CCFevents/shreve/abstracts/P.Carr.pdf.
[10] Derman, E. and Kani, I. (1994a). Riding on a smile, *RISK*, pp. 32–39.
[11] Dupire, B. (1994). Pricing with a smile, *Risk*, 7, pp. 18–20.
[12] Gatheral, J. (2006). The volatility surface: A practitionals guide. Wiley, Hoboken.

[13] Heston, S. (1993). A closed-form solution for options with stochastic volatility with applications to bond and currency options, *Review of Financial Studies*, 6, pp. 327–343.

[14] Itkin, A. (2015). To sigmoid-based functional description of the volatility smile, *North American Journal of Economics and Finance*, 31, pp. 264–291.

[15] Itkin, A. (2019). Deep learning calibration of option pricing models: some pitfalls and solutions, https://arxiv.org/abs/1906.03507,arxiv:1906.03507.

[16] Itkin, A. and Lipton, A. (2018). Filling the gaps smoothly, *Journal of Computational Sciences*, 24, pp. 195–208.

[17] Lipton, A. (2002). The vol smile problem, *Risk*, pp. 61–65.

[18] Lipton, A. and Sepp, A. (2011). Filling the gaps, *Risk Magazine*, pp. 86–91.

[19] Madan, D. and Seneta, E. (1990). The variance gamma (V.G.) model for share market returns, *Journal of Business*, 63(4), pp. 511–524.

[20] Merton, R. (1973). Theory of rational option pricing, Bell *Journal of Economics and Management Science*, 4, pp. 141–183.

[21] Merton, R. (1976). Option pricing when underlying stock returns are discontinuous, *Journal of Financial Economics*, 3, pp. 125–144.

[22] Samuelson, P. (1965). Rational theory of warrant pricing, *Industrial Management Review*, 6, pp. 13–32.

Preface

The concept of local volatility as well as the local volatility model is one of classical topics of mathematical finance. The model was invented independently by B. Dupire and E. Derman/I. Kani around 1994 as a relatively simple extension of the celebrated Black–Scholes model with the aim to explain the volatility smile and skew. They can be observed by comparing the implied volatilities of options written on the same underlying asset and having the same expiration date but different strikes. There exists an interesting opinion [Gatheral (2002)] that Dupire, Derman and Kani unlikely ever thought of local volatility as representing a model of how volatilities actually evolve. Rather, perhaps they introduced a notion of local volatilities as representing some kind of average over all possible instantaneous volatilities in a stochastic volatility world (an "effective theory"). As such, this model allows practitioners to price exotic options consistently with the known prices of vanilla options. This means that using the Dupire equation one can find a unique local volatility function $\sigma(K, T)$ from the market prices of standard options, and then construct, e.g., an implied tree that incorporates these local volatilities to value exotic options and to hedge standard options [Derman et al. (2016)]. Later this approach was further extended by introducing the local stochastic volatility models, see, e.g., [Bergomi (2016)] and references therein.

Despite, as mentioned, this topic with years has become classical and the existing literature on the subject is wide, yet there exist various problems with no sufficient attention drawn so far, e.g., (a) construction of analytical solutions of the Dupire equation for an arbitrary shape of the local volatility function, (b) construction of parametric or non-parametric regression of the local volatility surface suitable for fast calibration, (c) no-arbitrage interpolation and extrapolation of the local and implied volatility surfaces,

(d) extension of the local volatility concept beyond the Black-Scholes model, etc.

All these topics were of the author's interest at least within the last decade. Moreover, recent progress of deep learning and artificial neural networks as applied to financial engineering made it reasonable to look again at various classical problems of mathematical finance including that of building a no-arbitrage local/implied volatility surface and calibrating it to the option market data. Therefore, this book was written with the purpose of presenting our new results previously developed in a series of papers, and explaining them consistently starting from the general concept of Dupire, Derman and Kani and then being concentrated on various extensions proposed by the author and his co-authors. I felt the necessity to collect all the results in one place, also providing some typical examples of the problems that could be efficiently solved using the proposed methods. That was a motivation for this book to be finally written.

The book is conventionally split into three parts. In Part 1, we shortly describe the concept of the local volatility and the classical results of Dupire, Derman and Kani. Chapter 2 discusses no-arbitrage interpolation as this topic is important for all the remaining chapters, and is heavily used for our analytic constructions. We also shortly discuss some modern concepts of building the local volatility surface which are based on deep learning.

In Part 2, the local volatility model is considered as an extension of the Black–Scholes model. In Chapter 3, we consider a classical problem of calibration of the local volatility models to a given set of option prices. Here we present an extension of the approach proposed in [Lipton and Sepp (2011a)] which is developed by (a) replacing a piecewise constant local variance with a piecewise linear one, and (b) allowing non-zero interest rates and dividend yields. Our approach remains analytically tractable as it combines the Laplace transform in time with an analytical solution of the resulting spatial equations in terms of Kummer's degenerate hypergeometric functions. We also provide the results of various numerical experiments which demonstrate robustness of our approach. These results have been obtained in co-authorship with A. Lipton.

Chapter 4 presents an alternative view on this problem where construction of the local volatility surface is done by using a regression method, rather than the analytical solution of the Dupire equation. First, we give a short overview of the existing parameterizations, and also describe in more detail the SVI model of J. Gatheral which is popular among practitioners. We then describe another static parameterization of the implied volatility surface proposed by the author, which is constructed by using polynomials of sigmoid functions combined with some other terms. This parameterization is flexible enough to fit market implied volatilities which demonstrate smile or skew. An arbitrage-free calibration algorithm is considered that constructs the implied volatility surface as a grid in the strike-expiration space and guarantees a lack of arbitrage at every node of this grid. We also demonstrate how to construct an arbitrage-free interpolation and extrapolation in time, as well as build a local volatility and implied PDF surfaces. Asymptotic behavior of this parameterization is discussed, as well as results on stability of the calibrated parameters are presented. Numerical examples show the efficiency and of this approach in building these surfaces as well as demonstrate a better quality of the fit as compared with some known models.

Part 3 of the book is devoted to a new model which original version was invented by P. Carr in [Carr and Nadtochiy (2014)]). In that paper and later in [Carr and Nadtochiy (2017)], the concept of local volatility was applied to a time-homogeneous model where the underlying is driven by the arithmetic Brownian motion with stochastic time-change. The latter is done by using the Gamma process, so the model was called the Local Variance Gamma model. The original version relies on some restricted assumptions which include: no drift, a piecewise constant local volatility function, etc. However, in contrast to the approach in Chapter 3, this construction leads not to the Dupire PDE, but to a partial differential difference equation, which permits both explicit calibration and fast numerical valuation.

In Chapter 5, we describe an expanded version of this model by adding drift to the governing underlying process. Still in this new model it is possible to derive an ordinary differential equation for the option price which plays a role of Dupire's equation for the standard local volatility model. It is shown how calibration of multiple smiles (the whole local volatility surface) can be done in such a case. Further, assuming the local variance to be a piecewise linear function of strike and piecewise constant function of time this ODE is solved in closed form in terms of confluent hypergeometric functions. Calibration of the model to market smiles does not require solving any optimization problem and can be done term-by-term by solving a system of non-linear algebraic equations for each maturity, which is faster. These results are obtained in co-authorship with P. Carr.

In Chapter 6, we describe another extension of the Local Variance Gamma model. First, we develop a geometric version of this model with drift. Second, we consider three piecewise linear models: the local variance as a function of strike, the local variance as a function of log-strike, and the local volatility as a function of strike (so, the local variance is a piecewise quadratic function of strike). We show that for all these new constructions it is still possible to derive an ordinary differential equation for the option price, which plays a role of the Dupire equation for the standard local volatility model, and, moreover, it can be solved in closed form. Finally, similar to Chapter 5, we show that given multiple smiles the whole local variance/volatility surface can be recovered which does not require solving any optimization problem. These results are also obtained in co-authorship with P. Carr.

Overall, the book could potentially be helpful for those readers who want to learn of modern extensions of the local volatility model. Definitely, it doesn't pretend to give an extensive introduction into this subject, so it is assumed that the readers are already familiar with some basic concepts, e.g. by reading [Derman *et al.* (2016)]. Since from the mathematical point of view, the level of details is closer to the applied rather than to the abstract or pure theoretical mathematics, the book could also be recommended to graduate students with major in computational or quantitative finance, financial engineering or even applied mathematics.

I used to teach some topics of this book as a part of my special course on computational finance in School of Engineering of NYU in 2009–2019. I thank all my students for their questions, comments and remarks.

Andrey Itkin
Tandon School of Engineering
Department of Finance and Risk Engineering
New York University
New York, August, 2019

About the Author

Dr. Andrey Itkin is Director, Senior Quantitative Research Associate at Bank of America, and Adjunct Professor of Computational and Algorithmic Finance at the Department of Finance and Risk Engineering at Tandon School of Engineering, New York University. He holds a Ph.D. in Physics of Liquids, Gases, and Plasma from the Moscow Aviation University and a Doctor of Science degree in Computational and Molecular Physics from St. Petersburg Technical University. For a long time, he worked as a physicist, and then moved to finance as a quantitative researcher in equity derivatives, volatility trading, and market making, and then to risk management. He has authored numerous research papers in both physics and finance including the books *Microscopic Theory of Condensation in Gases and Plasma*, World Scientific, 1997, and *Pricing Derivatives under Levy models: Modern Finite Difference and Pseudo-differential Operators' Approach*, Springer, 2017. He is an Associate Editor of the *Journal of Derivatives* and the *International Journal of Computer Mathematics* and was a guest editor of various special issues including "Computational Methods in Finance" (*International Journal of Computer Mathematics*, Vol. 92, Issue 12, 2015), "Computational and Algorithmic Finance" (*Journal of Computational Science*, Vol. 24, 2018), and "Physics of Financial Derivatives" (*Journal of Derivatives*, 2019). He is also a member of multiple professional associations in finance and physics.

Acknowledgments

The idea of this book came from a series of papers written in co-authorship with Peter Carr and Alex Lipton. Therefore, it is my pleasure to acknowledge them, as many ideas described in this book came to my mind during our conversations and joint work. I am also grateful to my colleagues, Peter Friz, Dmitry Kreslavsky, Roza Galeeva, Antonie Kotzé, Lewis Biscamp, Archil Gulishashvili for various fruitful discussions. Last, but not least, I am thankful to the editor of this book, Ms. Yubing Zhai, who helped a lot to make this project being accomplished.

Contents

PART 1
General Concept of Local Volatility

Chapter 1

Local Volatility and Dupire's Equation

Local volatility model was invented around 1994 in [Dupire (1994)] for the continuous case and [Derman and Kani (1994a)] for the discrete case in response to the following problem.

In the celebrated Black-Scholes model, see e.g. [Hull (1997)], the dynamics of the stock price is modeled as a Geometric Brownian motion process with constant volatility parameter σ

$$dS_t = \mu S_t + \sigma S_t dW_t, \qquad S_t\Big|_{t=0} = S_0. \tag{1.1}$$

Here, S_t is the stock price at the time t, μ is the drift, W_t is the standard Brownian motion. It can be shown that under the risk-neutral measure $\mathbb{Q}$ the drift becomes $\mu = r - q$, where r, q are the constant interest rate and continuous dividends functions.

Next, we introduce a notion of the Black-Scholes implied volatility. In financial mathematics, the *implied volatility* σ_{BS} of an option contract is that value of the volatility of the underlying instrument which, when input in an Black-Scholes option pricing model will return a theoretical value equal to the current market price of the option. The implied volatility shows what the market implies about the underlying stock volatility in the future. For instance, the implied volatility is one of six inputs used in a simple option pricing (Black-Scholes) model, but is the only one that is not directly observable in the market. The standard way to determine it by knowing the market price of the contract and the other five parameters, is solving for the implied volatility by equating the model and market prices of the option contract. There exist various reasons why traders prefer considering option positions in term of the implied volatility, rather than the option price itself, see e.g., [Natenberg (1994)].

The Black-Scholes implied volatility is a useful measure, as it is a market practice instead of quoting the option premium in the relevant currency, the options are quoted in terms of the Black-Scholes implied volatility. Over the years, option traders have developed an intuition in this quantity. However, it can be further generalized by using a similar concept, but replacing the Black-Scholes framework with another one. For instance, in [Corcuera *et al.* (2009)] this is done under a Lévy framework, and, therefore, based on distributions that match more closely historical returns. Here we don't consider these generalizations, and are concentrated only on the Black-Scholes implied volatility.

Assume we are given an underlying and the continuous dividends function for this underlying. Also assume that the market constant interest rate is somehow known. Finally, assume that we are given a snapshot of market prices of, say European Call and Put options written on this underlying with the same expiration date T. Then one can compute the Black-Scholes implied volatilities for these options which will be a function of strikes K, and plot them against the strike price. Thus obtained line is called a *volatility smile* if it slopes upward on either end, or *volatility skew* if it slopes upward only on the left. The former behavior is typical for the stock options, while the latter — for the index options.

The important observation is that the volatility smiles should never occur based on standard Black-Scholes option theory, which normally requires a completely flat volatility curve. However, the first notable volatility smile was apparently seen back to 1987 following the stock market crash. Since that time the topic attracted a lot of attention in the financial industry. As mentioned in [Derman *et al.* (2016)], "After the crash, and ever since, equity index option markets have displayed a volatility smile, an anomaly in blatant disagreement with the Black-Scholes-Merton model. Since then, quants around the world have labored to extend the model to accommodate this anomaly". There exist various books on local and implied volatility with main focus on modeling, description, understanding, etc. We mentioned just few of them based on our own preferences, [Derman *et al.* (2016); Gatheral (2006); Natenberg (1994); Bossu (2014)], but the reader can also find numerous references therein.

The existence of the smile forced the quants to move from the simple Black-Scholes model to more sophisticated ones that would be able to describe this pattern. The idea of the *local volatility* model as proposed in [Dupire (1994); Derman and Kani (1994a)] was as follows. Assuming that only minimal changes should be applied to the Black-Scholes model, they proposed to replace the constant volatility σ with that which is a deterministic function of S_t and t. In other words, in this model instead of Eq.(1.31) we have

$$dS_t = \mu S_t + \sigma(S_t, t)S_t dW_t, \qquad S_t\Big|_{t=0} = S_0. \qquad (1.2)$$

Thus, in this model the local volatility is a function of the stock level S_t and time t (rather than the constant value) which might be sufficient to build a smile. With that, several natural questions become subject of an immediately concern [Derman *et al.* (2016)]:

(1) Is there exist a unique local volatility function or surface $\sigma(S, t)$ to match the observed implied volatility surface $\sigma_{BS}(S, t, K, T, r, q)$?

(2) If yes, that means that we can explain the observed smile by means of a local volatility process for the stock. Is the explanation meaningful? Does the stock actually evolve according to an observable local volatility function? There are many different models that can match the implied volatility surface, but achieving a match doesn't mean that model is "correct."

(3) What does the local volatility model tell us about the hedge ratios of vanilla options and the values of exotic options? How do the results differ from those of the classic Black-Scholes model?

The first question was positively answered by [Derman and Kani (1994a)] who constructed a binomial-tree model with volatilities at every model being calibrated to some stock and options market data. This modification of the binomial tree is called now the implied tree, [Hull (1997); Derman *et al.* (2016)]. In turn, [Dupire (1994)] considered the stock's local volatility function as $\sigma(K, T)$, i.e. this is the local volatility $\sigma(S, t)$ when the future stock price is K at time T. He derived a forward PDE (known now as the Dupire equation) which describes the dynamics of $\sigma(K, T)$ in the continuous case, which means the derivatives of $\sigma(K, T)$ represent the market prices of infinitesimal strike spreads, calendar spreads, and butterfly spreads. We discuss this derivation in the next Section.

1.1 Dupire's equation

The derivation of the Dupire equation is provided in many textbooks and papers, e.g., [Dupire (1994); Gatheral (2004); Derman *et al.* (2016); Rouah (2001)]. Here, we consider two ways to obtain it.

1.1.1 *Derivation using the Fokker-Planck equation*

Below suppose that the interest rate is deterministic, $r = r(t)$ and denote the discount factor $D(t, T)$ as

$$D(t, T) = \exp\left(-\int_t^T r(k)dk\right).$$ (1.3)

By definition, a European option Call $C(S, t, T)$ and Put $P(S, t, T)$ prices can be defined as, [Hull (1997)]

$$C(S, K, T - t) = D(t, T)\mathbb{E}_{\mathbb{Q}}[(S_T - K)^+],$$ (1.4)
$$P(S, K, T - t) = D(t, T)\mathbb{E}_{\mathbb{Q}}[(K - S_T)^+],$$

where for simplicity we dropped the dependence of the option prices on r, q, σ, $x^+ = \max(x, 0)$, and the expectation $\mathbb{E}_{\mathbb{Q}}$ is taken under the risk-neutral measure $\mathbb{Q}$. Therefore, it reads

$$\mathbb{E}_{\mathbb{Q}}[x] = \int xp(x, x_T, T - t)dx_T,$$

where $p(x, x_T, T - t)$ is the transition probability density from the state (x, t) into the state (x_T, T).

The function $p(S, S_T, T - t)$ satisfies the forward Kolmogorov (Fokker-Planck) equation (see, e.g. [Risken and Haken (1989); Soize (1994); Van Kampen (2007)] and references therein). With the simplified notation $p(S, t)$ instead of $p(S, S_T, T - t)$, it reads

$$\frac{\partial}{\partial t}p(S, t) = -\frac{\partial}{\partial S}[\mu Sp(S, t)] + \frac{1}{2}\frac{\partial^2}{\partial \sigma^2}[\sigma^2 S^2 p(S, t)].$$ (1.5)

This equation should be solved subject to the initial condition $p(S, 0) = \delta(S - S_0)$, where $\delta(x)$ is the Dirac delta-function.

The next step is to find an explicit expression for the option Theta: $\Theta = \frac{\partial C}{\partial T}$ by using the definition of C in Eq.(1.4). Using the chain rule we get

$$\frac{\partial C}{\partial T} = \frac{\partial D(t, T)}{\partial T}\int_K^\infty (S_T - K)p(S, S_T, T - t)dS_T$$ (1.6)

$$+ D(t, T)\int_K^\infty (S_T - K)\frac{\partial p(S, S_T, T - t)}{\partial T}dS_T$$

$$= -r(T)C + D(t, T)\int_K^\infty (S_T - K)\frac{\partial p(S, S_T, T - t)}{\partial T}dS_T.$$

Now the derivative under the integral in Eq.(1.7) can be substituted with the right hands side of Eq.(1.5) taken at $t = T$ which yields

$$\frac{\partial C}{\partial T} + r(T)C = D(t,T) \int_K^\infty (S_T - K)$$ (1.7)

$$\cdot \left\{ -\frac{\partial}{\partial S_T}[\mu(T)S_T p(S, S_T, T - t)] + \frac{1}{2}\frac{\partial^2}{\partial \sigma^2}[\sigma^2 S_T^2 p(S, S_T, T - t)] \right\} dS_T$$

$$= D(t,T) \left(-\mu(T)I_1 + \frac{1}{2}I_2 \right),$$

where the short notation I_1, I_2 is introduced for the integrals in the second line of Eq.(1.7). To evaluate these integrals we need two identities.

1.1.1.1 *First identity*

From the definition of the Call option price in Eq.(1.4) we have

$$\frac{C}{D(t,T)} = \int_K^\infty S_T p(S, S_T, T - t)dS_T - K\int_K^\infty p(S, S_T, T - t)dS_T.$$ (1.8)

On the other hand, from Eq.(1.4)

$$\frac{\partial C}{\partial K} = -D(t,T) \int_K^\infty p(S, S_T, T - t)dS_T.$$ (1.9)

Therefore,

$$\int_K^\infty S_T p(S, S_T, T - t)dS_T = \frac{1}{D(t,T)} \left(C - K\frac{\partial C}{\partial K} \right).$$ (1.10)

1.1.1.2 *The Breeden-Litzenberger identity*

Differentiating both sides of Eq.(1.9) on K and using the fundamental theorem of calculus, we obtain

$$\frac{\partial^2 C}{\partial K^2} = D(t,T)p(S, K, T - t).$$ (1.11)

This identity is known as the Breeden-Litzenberger formula, [Breeden and Litzenberger (1978)], which states that the risk-neutral probability of making a transition from S at time t to K at time T is proportional to the second partial derivative of the call price with respect to strike. ☐

With these identities we can proceed with evaluating the integrals I_1, I_2. For I_1 we get

$$I_1 = \int_K^\infty (S_T - K)\frac{\partial}{\partial S_T}[S_T p(S, S_T, T - t)]dS_T$$ (1.12)

$$= (S_T - K)\Big[S_T p(S, S_T, T - t)\Big]_K^\infty - \int_K^\infty S_T p(S, S_T, T - t) dS_T$$

$$= -\int_K^\infty S_T p(S, S_T, T - t) dS_T = \frac{1}{D(t,T)}\left(K\frac{\partial C}{\partial K} - C\right), \quad (1.13)$$

where in the last line the result in Eq.(1.10) is used, and also it is assumed that

$$\lim_{S_T \to \infty} (S_T - K) S_T p(S, S_T, T - t) = 0.$$

In other words, the first and the second moment of the density $p(S, S_T, T-t)$ are finite.

For I_2 we obtain

$$I_2 = \int_K^\infty (S_T - K)\frac{\partial^2}{\partial S_T^2}[\sigma^2 S_T^2 p(S, S_T, T - t)] dS_T \quad (1.14)$$

$$= (S_T - K)\frac{\partial}{\partial S_T}\Big[\sigma^2 S_T^2 p(S, S_T, T - t)\Big]_K^\infty$$

$$- \int_K^\infty \frac{\partial}{\partial S_T}[S_T p(S, S_T, T - t)] dS_T$$

$$= -\Big[\sigma^2 S_T^2 p(S, S_T, T - t)\Big]_K^\infty = \sigma^2 K^2 p(S, K, T - t).$$

Here $\sigma^2 = \sigma(K,T)^2$, and it is assumed that

$$\lim_{S_T \to \infty} (S_T - K)\frac{\partial}{\partial S_T}[S_T p(S, S_T, T - t)] = 0.$$

Using the Breeden-Litzenberger we finally obtain

$$I_2 = \frac{\sigma^2}{D(t,T)}K^2 \frac{\partial^2 C}{\partial K^2}. \quad (1.15)$$

1.1.1.3 *The final step*

Substituting Eq.(1.12) and Eq.(1.15) into Eq.(1.7) we obtain

$$\frac{\partial C}{\partial T} + r(T)C = \mu(T)C - \mu(T)K\frac{\partial C}{\partial K} + \frac{1}{2}\sigma^2 K^2 \frac{\partial^2 C}{\partial K^2}. \quad (1.16)$$

Taking into account that under the risk-neutral measure, the drift reads $\mu(T) = r(T) - q(T)$, [Brigo and Mercurio (2006)], we finally get the Dupire equation

$$\frac{\partial C}{\partial T} = \frac{1}{2}\sigma^2 K^2 - [(r(T) - q(T)]K\frac{\partial C}{\partial K} - q(T)C. \quad (1.17)$$

This equation can also be solved for the local variance $\sigma^2(K,T)$ to obtain

$$\sigma^2(K,T) = \frac{\frac{\partial C}{\partial T} + [(r(T) - q(T)]K\frac{\partial C}{\partial K} + q(T)C}{\frac{1}{2}K^2\frac{\partial^2 C}{\partial K^2}}. \tag{1.18}$$

1.1.2 *A probabilistic approach*

Here we derive the Dupire equation by using a probabilistic argument proposed in [Derman and Kani (1998)]. Let us define the stochastic variable

$$f(S_T, t, T) = D(t, T)(S_T - K)^+, \tag{1.19}$$

where S_t is the Geometric Brownian motion introduced in Eq.(1.31). Obviously, based on Eq.(1.4),

$$\mathbb{E}_{\mathbb{Q}}[f(S_T, t, T)|S_t = S] = C(S, K, T - t). \tag{1.20}$$

Using Itô's lemma at time $t = T$, one can find that $f(S_T, t, T)$ follows the process

$$df = \left(\frac{\partial f}{\partial T} + \mu(T)S_T\frac{\partial f}{\partial S_T} + \frac{1}{2}\sigma(T)^2S_T^2\frac{\partial^2 f}{\partial S_t^2}\right)dT + \sigma(T)S_T\frac{\partial f}{\partial S_T}dW_T. \tag{1.21}$$

The partial derivatives in this expression could be easily found using the definition of f in Eq.(1.19):

$$\frac{\partial f}{\partial T} = -r(T)D(t, T)(S_T - K)^+, \tag{1.22}$$

$$\frac{\partial f}{\partial S_T} = D(t, T)\mathbf{1}_{S_T > K}, \qquad \mathbf{1}_x = \begin{cases} 1, & x \geq 0 \\ 0, & x < 0 \end{cases}$$

$$\frac{\partial^2 f}{\partial S_T^2} = D(t, T)\delta(S_T - K).$$

Substituting this into Eq.(1.21) yields

$$df = D(t, T)\Big[\Big(-r(T)(S_T - K)^+ + \mu(T)S_T\mathbf{1}_{S_T > K} \tag{1.23}$$

$$+ \frac{1}{2}\sigma(T)^2S_T^2\delta(S_T - K)\Big)dT + \sigma(T)S_T\mathbf{1}_{S_T > K}dW_T\Big].$$

The first two terms in parentheses can be re-written as follows

$$-r(T)(S_T - K)^+ + \mu(T)S_T\mathbf{1}_{S_T > K} = \mathbf{1}_{S_T > K}[-r(T)(S_T - K) + \mu(T)S_T]$$

$$= r(T)K\mathbf{1}_{S_T > K} - q(T)S_T\mathbf{1}_{S_T > K}. \tag{1.24}$$

Since W_t is a martingale, taking the risk-neutral expectation of Eq.(1.23) and using Eq.(1.20) yields

$$dC = D(t,T)\mathbb{E}_\mathbb{Q}\Big[r(T)K\mathbf{1}_{S_T>K} - q(T)S_T\mathbf{1}_{S_T>K} \qquad (1.25)$$
$$+ \frac{1}{2}\sigma(T)^2 S_T^2 \delta(S_T - K)\Big]dT.$$

Since

$$D(t,T)\mathbb{E}_\mathbb{Q}[S_T\mathbf{1}_{S_T>K}] = C + KD(t,T)\mathbb{E}_\mathbb{Q}[\mathbf{1}_{S_T>K}],$$

and

$$\frac{\partial C}{\partial K} = -D(t,T)\mathbb{E}_\mathbb{Q}[\mathbf{1}_{S_T>K}]. \qquad (1.26)$$

Eq.(1.25) can also be represented as

$$\frac{\partial C}{\partial T} = D(t,T)K[r(T) - q(T)]\mathbb{E}_\mathbb{Q}[\mathbf{1}_{S_T>K}] - q(T)C \qquad (1.27)$$
$$+ \frac{1}{2}D(t,T)\mathbb{E}_\mathbb{Q}[\sigma^2(T)S_T^2\delta(S_T - K)]$$
$$= -K[r(T) - q(T)]\frac{\partial C}{\partial K} - q(T)C + \frac{1}{2}D(t,T)\mathbb{E}_\mathbb{Q}[\sigma^2(T)S_T^2\delta(S_T - K)].$$

The last term in this equation can be simplified by using the sifting property of the Dirac delta function

$$\frac{1}{2}D(t,T)\mathbb{E}_\mathbb{Q}[\sigma^2(T)S_T^2\delta(S_T - K)] = \frac{1}{2}D(t,T)\mathbb{E}_\mathbb{Q}[\sigma^2(T)S_T^2|S_T = K] \quad (1.28)$$
$$\cdot \mathbb{E}_\mathbb{Q}[\delta(S_T - K)] = \frac{1}{2}D(t,T)\mathbb{E}_\mathbb{Q}[\sigma^2(T)|S_T = K]K^2\mathbb{E}_\mathbb{Q}[\delta(S_T - K)]$$
$$= \frac{1}{2}\mathbb{E}_\mathbb{Q}[\sigma^2(T)|S_T = K]K^2\frac{\partial^2 C}{\partial K^2},$$

since it follows from Eq.(1.26) that

$$\frac{\partial^2 C}{\partial K^2} = D(t,T)\mathbb{E}_\mathbb{Q}[\delta(S_t - K)].$$

Thus, from Eq.(1.27) and Eq.(1.28) we finally obtain

$$\frac{\partial C}{\partial T} = -K[r(T) - q(T)]\frac{\partial C}{\partial K} - q(T)C + \frac{1}{2}\mathbb{E}_\mathbb{Q}[\sigma^2(T)|S_T = K]K^2\frac{\partial^2 C}{\partial K^2}. \qquad (1.29)$$

Comparing this with the Dupire equation Eq.(1.17) we see that

$$\sigma^2(K,T) = \mathbb{E}_\mathbb{Q}[\sigma^2(S_T,T)|S_T = K].$$

This means that the local variance is the risk-neutral expectation of the instantaneous variance conditional on the final stock price S_T being equal to the strike price K.

1.2 Local volatility via the implied volatility

In this Section we derive the identity that connects the local and Black-Scholes implied variances. The identity was introduced in [Lipton (2002); Gatheral (2006)] and reads

$$\sigma^2(T,K) = \frac{\partial_T w}{\left(1 - \frac{y \partial_y w}{2w}\right)^2 - \frac{(\partial_y w)^2}{4}\left(\frac{1}{w} + \frac{1}{4}\right) + \frac{\partial_y^2 w}{2}}, \qquad (1.30)$$

where $y = \log K/F$, $F = Se^{(r-q)T}$ is the stock forward price, and $w = \sigma_{BS}^2 T$ is the total implied variance. In terms of these variables the Black-Scholes formula for the future value of the Call option price becomes, [Gatheral (2006)]

$$C_{BS}(F_T, y, w) = D(t,T)F_T[N(d_1) - e^y N(d_2)] \qquad (1.31)$$

$$= D(t,T)F_T\left[N\left(-\frac{y}{\sqrt{w}} + \frac{\sqrt{w}}{2}\right) - e^y N\left(-\frac{y}{\sqrt{w}} - \frac{\sqrt{w}}{2}\right)\right],$$

where $N(x)$ is the normal CDF, and

$$d_1 = \frac{\ln\frac{F_T}{K} + \frac{\sigma_{BS}^2}{2}T}{\sigma_{BS}\sqrt{T}}, \qquad d_2 = d_1 - \frac{1}{2}\sigma_{BS}\sqrt{T}. \qquad (1.32)$$

The Dupire equation Eq.(1.17) can also be re-written in terms of the new variables. It is easy to check that this yields

$$\frac{\partial C}{\partial T} = \frac{1}{2}\sigma^2\left[\frac{\partial^2 C}{\partial y^2} - \frac{\partial C}{\partial y}\right] - qC. \qquad (1.33)$$

Computing derivatives of the Black-Scholes formula in Eq.(1.31), we obtain

$$\frac{\partial^2 C_{BS}}{\partial w^2} = \left(-\frac{1}{8} - \frac{1}{2w} + -\frac{y^2}{2w^2}\right)\frac{\partial C_{BS}}{\partial w}, \qquad (1.34)$$

$$\frac{\partial^2 C_{BS}}{\partial w \partial y} = \left(\frac{1}{2} - \frac{y}{w}\right)\frac{\partial C_{BS}}{\partial w},$$

$$\frac{\partial^2 C_{BS}}{\partial y^2} - \frac{\partial C_{BS}}{\partial y} = 2\frac{\partial C_{BS}}{\partial w}.$$

Taking into account that $w = w(y, T)$, the Dupire equation in Eq.(1.33) can be also transformed by using the identities

$$\frac{\partial C}{\partial y} = \frac{\partial C_{BS}}{\partial y} + \frac{\partial C_{BS}}{\partial w}\frac{\partial w}{\partial y}, \tag{1.35}$$

$$\frac{\partial^2 C}{\partial y^2} = \frac{\partial^2 C_{BS}}{\partial y^2} + 2\frac{\partial^2 C_{BS}}{\partial w \partial y}\frac{\partial w}{\partial y} + \frac{\partial^2 C_{BS}}{\partial w^2}\left(\frac{\partial w}{\partial y}\right)^2 + \frac{\partial C_{BS}}{\partial w}\frac{\partial^2 w}{\partial y^2},$$

$$\frac{\partial C}{\partial T} = \frac{\partial C_{BS}}{\partial T} + \frac{\partial C_{BS}}{\partial w}\frac{\partial w}{\partial T} = \frac{\partial C_{BS}}{\partial w}\frac{\partial w}{\partial T} - qC_{BS}.$$

Substituting these expressions into Eq.(1.33) yields

$$\frac{\partial C_{BS}}{\partial w}\frac{\partial w}{\partial T} = \frac{1}{2}\sigma^2\frac{\partial C_{BS}}{\partial w}A, \tag{1.36}$$

$$A = 2 - \frac{\partial w}{\partial y} + \left(-\frac{1}{8} - \frac{1}{2w} + \frac{y^2}{2w^2}\right)\left(\frac{\partial w}{\partial y}\right)^2 + \frac{\partial^2 w}{\partial y^2}$$

$$+ 2\left(\frac{1}{2} - \frac{\partial y}{\partial w}\right)\frac{\partial w}{\partial y}.$$

Taking out a factor $\frac{\partial C_{BS}}{\partial w}$ and simplifying, we finally obtain Eq.(1.30).

An interesting particular case is when $\frac{\partial w}{\partial y} = 0$. This implies $\frac{\partial \sigma_{BS}(S,K,T)}{\partial K} = 0$, i.e., there is no skew and the implied volatility is flat. Then it follows from Eq.(1.30) that $\sigma^2 = \frac{\partial w}{\partial T}$. In other words, in this case the local variance reduces to the forward Black-Scholes implied variance, and

$$w(T) = \int_0^T \sigma^2(k)dk. \tag{1.37}$$

Also, it is possible to show that the implied volatility $\sigma_{BS}(S, K, T)$ of an option is approximately the average of the local volatilities $\sigma(S, t)$ encountered over the life of the option between the current underlying price and the strike. A detailed discussion of this rule of thumb is provided in [Derman *et al.* (2016)].

Chapter 2

Local Volatility Surface and No-arbitrage

Suppose that the local volatility function $\sigma(K,T)$ is somehow known. Also, suppose that given $\sigma(K,T)$, there exist theoretical option prices that solve the Dupire equation derived in the previous chapter. Finally, suppose that for a given set of option strikes K and maturities T these theoretical prices exactly coincide with the corresponding market prices. Then one can build a local volatility surface by using the values of $\sigma(K,T)$ at the given set of $[K,T]$, and say that this surface represents the given set of the market quotes. Note, that in practice, we usually solve the inverse problem. This is, given the market quotes, find the local volatility function that being used in the Dupire equation produces the theoretical option prices such that the difference between this set of option prices and the corresponding market prices reaches minimum under some suitable norm.

It is well-known, that the theoretical option prices obtained by solving the Dupire equation are no-arbitrage (we will discuss this notion below in this chapter). Therefore, if build the local volatility surface as described in above, at the given set of points $[K,T]$ our surface also respects no-arbitrage. This is true even if the market prices are arbitragable. Indeed, in the latter case the theoretical prices found by solving the Dupire equation, are no-arbitragable themselves, but will deviate from those at the market.

Suppose, however, that in addition to the option prices at the given $[K_i,T_j]$, $i,j \in \Omega$, where Ω is a set of the discrete pairs $[i,j]$, $i = 1,...,N$, $j = 1,...,M$, corresponding to the given market quotes and used to build the local volatility surface, we also need those for $[i,j] \notin \Omega$. An immediate idea is to use some kind of interpolation. However, in this case we need to be careful about no-arbitrage, as, by default, interpolation doesn't preserve it. Therefore, in this chapter we discuss construction of a no-arbitrage local volatility surface. The proposed constructions are then widely used in the book for several purposes. First, when building tractable analytical methods for solving Dupire's or Dupire-like equations, no-arbitrage interpolation helps with computing the killing terms of the equation in closed from. Second, even when regression methods such as those described in Chapter 4 are in use, still the no-arbitrage conditions must be taken into account at every point of the surface to guarantee no-arbitrage at least at this points. While the details of this are discussed in the corresponding chapters of this book, here we present a general description of the problem and provide some helpful solutions developed by the author in [Itkin and Lipton (2018); Carr and Itkin (2018a,b)].

2.1   No-arbitrage conditions and interpolation

According to [Cox and Rubinstein (1985)][1], given three Put option prices $P(K_1), P(K_2), P(K_3)$ for three strikes $K_1 < K_2 < K_3$, the necessary and sufficient conditions for those prices to be arbitrage-free, read

$$P(K_3) > 0, \qquad P(K_2) < P(K_3), \tag{2.1}$$
$$(K_3 - K_2)P(K_1) - (K_3 - K_1)P(K_2) + (K_2 - K_1)P(K_3) > 0.$$

Suppose that we want to use linear interpolation in the strike space on the interval $[K_1, K_3]$ to find the unknown Put option price $P(K_2)$ given the values of $P(K_1), P(K_3)$,

$$P(K_2) \equiv P_l(K_2) = \frac{P(K_1)K_3 - P(K_3)K_1}{K_3 - K_1} + \frac{P(K_3) - P(K_1)}{K_3 - K_1}K_2.$$

When plugging this expression into the second line of Eq.(2.1), the LHS of the latter vanishes, so the third no-arbitrage condition is violated.

This problem, however, could be resolved if we use linear interpolation with a modified independent variable,

$$P(K_2) \equiv P_F(K_2) \tag{2.2}$$
$$= \frac{P(K_1)f(K_3) - P(K_3)f(K_1)}{f(K_3) - f(K_1)} + \frac{P(K_3) - P(K_1)}{f(K_3) - f(K_1)}f(K_2),$$

where $f(K)$ is a convex and increasing function in $[K_1, K_3]$. Indeed, if $f(K)$ is convex, then $P(K_2) = P_F(K_2) = P_l(K_2) - \varepsilon$, $\varepsilon > 0$ (see Fig. 2.1). Substitution of this expression into the second line of Eq.(2.1) gives

---

[1]In [Cox and Rubinstein (1985)] these conditions are given for Call option prices. In that case the first and the third conditions remain the same as in Eq.(2.1) if we replace $P(K)$ with $C(K)$, while the second condition changes to $C(K_2) > C(K_3)$.

$(K_3 - K_1)\varepsilon > 0$, which is true. The second condition in Eq.(2.1) now reads

$$(P(K_1) - P(K_3))(f(K_3) - f(K_2))(f(K_1) - f(K_3)) > 0,$$

which is also true since $f(K)$ is an increasing function of K.

Alternatively, one can use non-linear interpolation. For instance, in [Itkin and Lipton (2018)] by combining both approaches for the sake of tractability, the following interpolation scheme is proposed

$$P(K_2) \equiv P_F(K_2) = \gamma_1 + \gamma_2 K_2 \log K_2, \qquad (2.3)$$
$$\gamma_1 = \frac{P_3 K_1 \log K_1 - P_1 K_3 \log K_3}{K_1 \log K_1 - K_3 \log K_3},$$
$$\gamma_2 = \frac{P_1 - P_3}{K_1 \log K_1 - K_3 \log K_3}.$$

Proposition 2.1. *The interpolation scheme in Eq.(2.3) preserves no-arbitrage.*

Proof. Observe, that the no-arbitrage conditions in Eq.(2.1) are discrete versions of the conditions

$$P > 0, \quad P_K > 0, \quad P_{K,K} > 0.$$

By differentiating the first line of Eq.(2.3) one can check that the proposed interpolation obeys these conditions provided that P is an increasing function of K given the values of all other parameters to be constant. For instance, this is the case for the Black-Scholes Puts.

At very small K_2 the derivative P_K in Eq.(2.3) is still positive, because γ_2 in this limit becomes negative and tends to zero. $\qquad \square$

For the sake of illustration, in Fig. 2.1 we present a comparison of the no-arbitrage interpolation P_N with its linear counterpart P_L and the exact price P_E computed for the Black-Scholes model (for emphasis, the differences $D(P_N) = P_N - P_L, D(P_E) = P_E - P_L$ are displayed). The plot is computed using the following values: $S = 100, K_1 = 95, K_3 = 100, r = 0.05$, $q = 0.01, \sigma = 0.5, T = 1$. It is clear that the no-arbitrage conditions are satisfied.

2.2 No-arbitrage at consecutive intervals

Proposition 2.1 guarantees that the proposed interpolation doesn't introduce an arbitrage into the solution if any three strikes belong to the same interval $[K_1, K_3]$. However, what if we consider strikes K_2, K_3, K_4 as this is schematically depicted in Fig. 2.2.

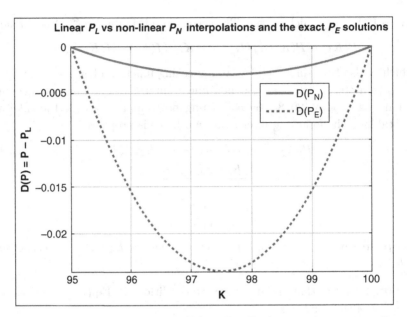

Figure 2.1: Absolute differences $D(P) = P - P_L$ for no-arbitrage non-linear interpolation P_N, and the exact Black-Scholes Put prices P_E, with the linear interpolation P_L. The line $D(P_L) = 0$ corresponds to P_L.

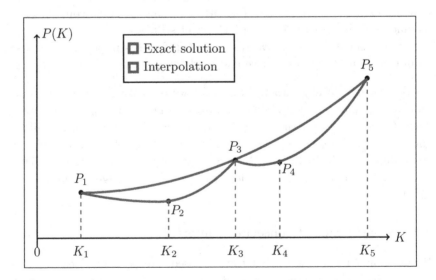

Figure 2.2: Three strikes K_2, K_3, K_4 which belong to the consecutive intervals.

Here at the interval $[K_1, K_3]$ the exact solution is depicted by the red line, while our quadratic interpolation is in blue. Accordingly, the Put prices P_1, P_3 are the market quotes, so they assumed to be the exact prices with no market arbitrage. By our construction, these prices also don't have a model arbitrage. At the consecutive interval $[K_3, K_5]$ a similar construction applies.

We have to emphasize that this graph is pure illustrative, and no-arbitrage interpolation guarantees that $P'(K) > 0$, while the blue line in Fig. 2.2 doesn't support this. However, if we draw an accurate picture by using the above formulae, it would be almost impossible to distinguish the red and blue lines. Therefore, we changed convexity and skew of the blue line to make the difference visible.

By Proposition 2.1 given a set of strikes K_1, K_2, K_3 the price P_2 obtained by interpolation preserves no-arbitrage. The same is true for P_5 given the Put prices P_3, P_5 at strikes K_3, K_5. Now assume that given K_1, K_3, K_5 and P_1, P_3, P_5 we want to check the no-arbitrage conditions for the set of strikes K_2, K_3, K_4. The Proposition 2.1 doesn't help in this situation, so we need a special consideration of this case.

Obviously, the first and second conditions in Eq.(2.1) are still satisfied in this case, so we need to check that the butterfly spread is positive. Unfortunately, at the moment we don't have a general analytical solution of this problem, while some particular cases can be addressed. Thus this remains an open question. However, we checked this condition numerically. In doing so we used the Black-Scholes Put prices P_1, P_3, P_5[2] and built a 2D plot of Bs which is the left-hands side of the third line in Eq.(2.1). The results for two cases presented in Table 2.1 are presented in Figs. 2.3 and 2.4.

Table 2.1: Parameters of the test for non-negativity of the Butterfly spread. σ_{BS} is the Black-Scholes implied volatility.

Test	S	r	σ_{BS}	T	K_1	K_3	K_5
1	100	0.01	0.5	2	80	100	130
2	100	0.1	0.1	0.1	90	100	105

Overall, we ran a lot of tests and didn't find any case where the butterfly spread would become negative. This partly supports our no-arbitrage interpolation. More sophisticated cases where, e.g., instead of strike K_3 in

---

[2]This is done to preserve upper bounds on the Put price that $P(S, K, T, r) \leq Ke^{-rT}$, [Levy (1985)].

the butterfly spread at strikes K_2, K_3, K_4 we use another strike K_6 such that $K_1 < K_2 < K_3 < K_4 < K_6 < K_5$, could be treated in a similar way. Again, our numerical tests didn't reveal any case where a butterfly spread would become negative.

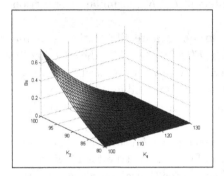

Figure 2.3: Butterfly spread Bs for a set of strikes K_2, K_3, K_4 computed in Test 1 in Table 2.1.

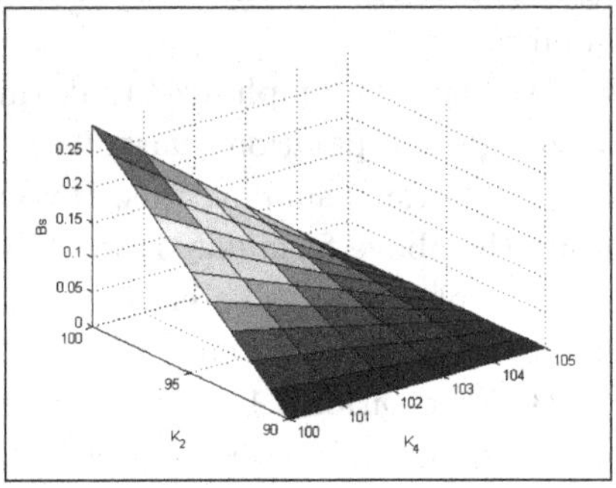

Figure 2.4: Butterfly spread Bs for a set of strikes K_2, K_3, K_4 computed in Test 2 in Table 2.1.

2.3   Another types of no-arbitrage interpolation

As shown in Section 2.1, to construct a no-arbitrage interpolation one can use a suitable non-linear interpolation. The non-linear function, however, is not unique. It could be chosen based on some additional consideration. For instance, the interpolation scheme in Eq.(2.3) is further used in Section 3.3, see also [Itkin and Lipton (2018)]). With this representation the modified Put price (which is a dependent variable of the approach) acquires a nice tractable representation, so the integral I_{12} can be computed in closed form. Here we want to exploit the same idea, but propose another interpolation scheme which will be used in Chapter 5.

Below let us consider the following interpolation scheme

$$P(x) \equiv P_F(x) = \gamma_1 + \gamma_2 x^2, \quad x_1 \le x \le x_3, \tag{2.4}$$
$$\gamma_1 = \frac{P(x_3)x_1^2 - P(x_1)x_3^2}{x_1^2 - x_3^2}, \qquad \gamma_2 = \frac{P(x_1) - P(x_3)}{x_1^2 - x_3^2}.$$

Then similar to Proposition 2.1, we can prove the following Proposition

Proposition 2.2. *The interpolation scheme in Eq.(2.3) is arbitrage free at the interval* $[K_1, K_3]$.

Proof. Observe, that the no-arbitrage conditions in Eq.(2.1) are discrete versions of the conditions

$$P > 0, \quad P_K > 0, \quad P_{K,K} > 0.$$

They, in turn, correspond to the conditions

$$P > 0, \quad P_x > 0, \quad P_{x,x} > 0,$$

as $x'(K) = 1/S > 0$. By differentiating the first line of Eq.(2.4) one can check that the proposed interpolation obeys these conditions provided that P is an increasing function of K (or x) given the values of all other parameters to be constant. For instance, this is the case for the Black-Scholes Puts. □

Note, that the concept of no-arbitrage interpolation is widely used throughout this book. In Chapter 6 we describe the Geometric Local Variance Gamma model introduced in [Carr and Itkin (2018b)]. Three particular models of the local variance function are considered there: the local variance as a piecewise linear function of strike, the local variance as a piecewise linear function of log-strike, and the local volatility as a piecewise linear function of strike (so, the local variance is a *piecewise quadratic* function of strike). We show that in this new model it is still possible to derive an ordinary differential equation for the option price, which plays a role of Dupire's equation for the standard local volatility model. Moreover, it all three cases, this equation can be solved in closed form. This is achieved by replacing the continuous function — the local variance in the killing term — with its no-arbitrage interpolator. Again, as there exist many no-arbitrage interpolations, a particular form is chose in such a way to make all the integrals to be taken in closed form. These interpolations are discussed in detail in Section 6.4.

In Section 4.3.6 we discuss a no-arbitrage interpolation of the implied volatility surface. This interpolation is inspired by the approach [Gatheral and Jacquier (2014)], despite differs in many details. This interpolation is not focused on tractability as the model is discrete and calibrated numerically, however is useful as provides a natural financial meaning of the construction.

2.4 No-arbitrage as positivity

In the Black-Scholes world the no-arbitrage conditions which we discussed in the previous Sections are equivalent to the existence of a risk-neutral density (state price density). This density allows finding the option price in every possible state and is connected to the second derivative of the option price on strike by the Breeden-Litzenberger formula Eq.(1.11).

Rebonato in [Rebonato (2004)] discusses the conditions that guarantee the existence of a risk neutral density, which finally give rise to Eq.(2.1). He separates them into the following categories:

(1) Market Conditions: The market is complete, frictionless, there are no bid-ask spreads, short sales are allowed and there are nt taxes.
(2) Traded Instruments: There exist and traded the underlying asset and plain-vanilla calls and puts options for all maturities and strikes. There also exist bonds with the payoff determined by a risk free interest.
(3) Probability Spaces: Market information is determined by using a filtered probability space $(\Omega, \mathcal{F}_t, \mathbb{Q})$ where Ω is the state space, $\mathcal{F}_t$ is the filtration and $\mathbb{Q}$ is the risk-neutral probability measure. The state space Ω contains all present and possible future values of the underlying asset and derivative options, and $\mathcal{F}_t$ is the natural filtration generated by history of prices of the underlying and the options for a finite but large number of dates.

The no-arbitrage conditions Eq.(2.1) are a useful alternative to the positivity of the implied density as they can be easily checked having a set of option prices for the same underlying but various strikes and maturities. However, as this could be see below in Chapter 4, these conditions require either various restrictions on parameters of the implied volatility surface, or running the constrained optimization which is computationally expensive. Therefore, in [Carr and Pelts (2015)] the authors ask a reasonable question: might there be an alternative to an implied variance rate surface for which no arbitrage just requires positivity?

Indeed, in the local volatility model one needs a positive local volatility function $\sigma(K, T)$ to generate an arbitrage-free implied volatility surface. In [Schweizer and Wissel (2008)] for getting this surface the authors quote a positive function of just moneyness $M = S/K$. And in [Carr and Pelts (2015)] the authors claim the need of either one positive function of log-moneyness, or one positive function of log-moneyness and one positive function of maturity. In the second case the positive function of T controls the level of ATM implied volatility at each term, and the second positive function of M controls the implied volatility skew across all terms. Below we shortly present the approach of [Carr and Pelts (2015)] (note also a recent extension of this approach in [Antonov et al. (2019)]) as it demonstrates a useful alternative way of constructing a no-arbitrage implied (and so local) volatility surface by expressing it not in terms of the option prices or implied variance rates, but some alternatives.

So, Carr and Pelts start with considering the FX options with the intrinsic value given by $(N_+ - N_-)^+$, where N_+ denotes the spot price in the specified pricing currency of the contract received in the optional exchange, and N_- denotes the spot price in the same pricing currency of the contract delivered. The option price is given by the pricing function $P(N_+, N_-, T) : \mathbb{R}^+ \times \mathbb{R}^+ \times \mathbb{R}^+ \mapsto \mathbb{R}^+$. The ultimate goal is to find an unconstrained non-negative alternative to $P(N_+, N_-, T)$, which respects all of no arbitrage constraints.

The function $P(N_+, N_-, T)$ is linearly homogeneous in the first two arguments, and, therefore, follows the Euler representation

$$P(N_+, N_-, T) = N_+ P_1(N_+, N_-, T), + N_- P_2(N_+, N_-, T), \qquad (2.5)$$

where P_1, P_2 are first partial derivatives of $P(N_+, N_-, T)$ with respect to N_+ and N_-, respectively. They can be referred as options Deltas: Δ_+, Δ_-, and also as probabilities of the option finishing in-the-money under the corresponding measures $\mathbb{Q}_+$ and $\mathbb{Q}_-$:

$$\Delta_+ = P_1(N_+, N_-, T) = \mathbb{Q}_+\Big\{N_{+,T} > N_{-,T}\Big\}, \qquad (2.6)$$

$$\Delta_- = -P_2(N_+, N_-, T) = \mathbb{Q}_-\Big\{N_{+,T} > N_{-,T}\Big\}.$$

Let's now introduce

$$\hat{P}(R, T) = \frac{P(N_+, N_-, T)}{N_-}, \quad R = \frac{N_+}{N_-}. \qquad (2.7)$$

From Eq.(2.5) and Eq.(2.7) we have

$$-P_2(N_+, N_-, T) = R\hat{P}_1(R, T) - \hat{P}(R, T), \qquad (2.8)$$

$$\hat{P}_1(R, T) \equiv P_1(N_+, N_-, T).$$

The right hands part of the first line in Eq.(2.8) can be written as a function of $\Delta_+ = P_1(N_+, N_-, T) = \hat{P}_1(R, T)$ in the form

$$\mathcal{D}_-(\Delta_+, T) = R\hat{P}_1(R, T) - \hat{P}(R, T) = \sup_{R>0}\Big[R\Delta_+ - \hat{P}(R, T)\Big]. \qquad (2.9)$$

Then the function $\mathcal{D}_-(\Delta_+, T)$ becomes the Legendre transform of the convex function $\hat{P}(R, T)$, see, e.g., [Carr (2014a); Itkin (2018)]. It can be verified that $\forall T > 0$ $\mathcal{D}_- : [0, 1] \mapsto [0, 1]$, $\mathcal{D}_-(0, T) \mapsto 0$, $\mathcal{D}_-(1, T) \mapsto 1$, and $\mathcal{D}_-(\Delta_+, T)$ is an increasing function of Δ_+. Therefore, $\mathcal{D}_-(\Delta_+, T)$ is the distortion function, [Denneberg (1990)].

An important observation made by [Carr and Pelts (2015)] is that if a convex distortion function $\Delta_- = \mathcal{D}_-(\Delta_+, T)$ is somehow specified directly $\forall T > 0$, it also generates the convex function linking the normalized option price $\hat{P}_1(R, T)$ to the ratio R of its underlyings

$$\hat{P}_1(R, T) = \sup_{\Delta_+ \in [0,1]} [R\Delta_+ - \mathcal{D}_-(\Delta_+, T)] \qquad (2.10)$$

$$= R\Delta_+(R, T) - \mathcal{D}_-(\Delta_+(R, T), T),$$

where $\Delta_+(R, T)$ is the inverse of the increasing function $R = \frac{\partial \mathcal{D}_-}{\partial \Delta_+}(\Delta_+, T)$. Moreover, the convexity of Δ_- in Δ_+ implies the convexity of $\hat{P}$ in R and so in N_+ and N_-. Therefore, such the option pricing function will be free of butterfly spread arbitrage.

To avoid the arbitrage in T (a calendar spread arbitrage) $\forall R \geq 0$, the normalized option price $\hat{P}$ must be increasing in T. It turns out that this is equivalent to $\mathcal{D}_-$ be the decreasing function of T also $\forall \Delta_+ \in [0.1]$. Thus, to provide no arbitrage of all types the distortion function should be convex in $\Delta_+ \in [0.1]$ and increasing in $T \geq 0$.

This statement is the main result of [Carr and Pelts (2015)]. Then they discuss various ways of generating an arbitrage-free distortion function. For instance, for the Black-Scholes model the stochastic variable $\log R_T$ is normally distributed under both measures $\mathbb{Q}_+$ and $\mathbb{Q}_-$. It is well-known that $\Delta_- = N(-d_1)$, $\Delta_+ = N(-d_2)$ (see Eq.(1.32)), and hence the distortion function is

$$\Delta_- = N\left(N^{-1}(\Delta_+) - \sqrt{\sigma^2 T}\right). \qquad (2.11)$$

This type of distortion function is called a Wang transform. It can be checked that this distortion function is arbitrage-free, i.e. it obeys the conditions specified in the previous paragraph.

The second step in [Carr and Pelts (2015)] is to generalize this approach of constructing an arbitrage-free distortion function for some non-Black Scholes models. For doing that they choose two variables z_+, z_- to represent log-moneyness, similar to what $-d_2$ and $-d_1$ variables do in the Black-Scholes world. Then they introduce a generalized Wang Transform (compare with Eq.(2.11)

$$\Delta_- = \Omega\left(\Omega^{-1}(\Delta_+) - \tau(T)\right). \qquad (2.12)$$

Here Ω is a CDF of some random variable $Z_T \in \mathbb{R}$, and $\tau(T) \geq 0$. For convexity of Δ_- we need that the function $R(\Delta_+, T) = \frac{\partial \Delta_-}{\partial \Delta_+}$ is increasing in Δ_+. This could be achieved, e.g., if both R and Δ_+ are increasing functions of some third variable $z_- \in \mathbb{R}$ at each $T \geq 0$. Assuming $z_- = \Omega^{-1}(\Delta_-)$, from Eq.(2.12) we obtain $\Delta_+ = \Omega(z_- + \tau(T))$, i.e. indeed Δ_+ is increasing in z_-.

Both parts of Eq.(2.12) can be differentiated with respect to Δ_+ to obtain

$$R = \frac{\Omega'(z_-)}{\Omega'(z_+)}, \qquad z_+ = z_- + \tau(T).$$

Then Carr and Pelts show that a sufficient condition for R to be increasing in z_- at each $T \geq 0$ is that the PDF $\Omega'(z_-)$ is log concave in z_-. Then the generalized Wang Transform defined in Eq.(2.12) is an arbitrage-free distortion function.

To construct such a function, they also observe that the log concavity of $\Omega'(z_-)$ is equivalent to the convexity of another function $h(z_-) = -\log \Omega'(z_-)$ in z_-. And a convex function $h(z_-) : \mathbb{R} \mapsto \mathbb{R}$ can be generated by picking a positive function $p(z_-) : \mathbb{R} \mapsto \mathbb{R}^+$ and then integrating it twice in z_-. In a similar way, an increasing function $\tau(T) : \mathbb{R}^+ \mapsto \mathbb{R}^+$, $\tau(o) = 0$ can be constructed by picking a positive function $q(T) : \mathbb{R}^+ \mapsto \mathbb{R}^+$ and integrating it once in T.

In the simplest case when $p(z_-) = p_0 > 0$, i.e. constant, $h(z_-)$ is a quadratic function, and the log concave PDF $\Omega'(z_-) = e^{-h(z_-)}$ is Gaussian. Furthermore, $\log R_T$ is linear in Z_T, so is also Gaussian. Hence, flat $p(z_-)$ implies the Black Scholes variance as does flat implied variance. However, the former is arbitrage-free while the latter can produce vertical and/or butterfly spread arbitrage.

PART 2
The Black-Scholes Model

Chapter 3

Analytical Methods of Building the Local Volatility Surface

In Chapter 1 a general concept of the local volatility model invented by [Dupire (1994)] and [Derman and Kani (1994a)], and some basic definitions and notions have been introduced. This chapter continues studying this model in more detail. Throughout the chapter we deal with a classical flavor if the local volatility model where the underlying stochastic process is represented by the Geometric Brownian Motion with the constant volatility replaced by a local volatility function. The other model will be discussed in Part 3 of this book.

Our focus here is one of the classical problems of mathematical finance: given a current snapshot of option prices for various strikes and maturities written on the same underlying, build a local volatility surface which in a certain norm minimizes the difference between the market quotes and the corresponding values from the surface. Solving this problem is called *calibration*.

So far we didn't speak about calibration. The calibration of the local volatility (LV) surface to the market data, representing either prices of European options or the corresponding implied volatilities for a given set of strikes and maturities, drew a lot of attention over the past two decades. Various approaches to solving this important problem were proposed, see, e.g., [Andreasen and Huge (2011); Lipton and Sepp (2011b); Itkin (2015)] and references therein. Below, we refer to [Lipton and Sepp (2011b)] as LS2011 for the sake of brevity.[1]

There are two main approaches to solving the calibration problem. The first approach attempts to construct a continuous implied volatility (IV) surface matching the market quotes by using either some parametric or

---

[1]We emphasize that the solution proposed in [Andreasen and Huge (2011)] is static in nature, while the solution developed in LS2011 is fully dynamic.

non-parametric regression, and then generates the corresponding LV surface via the Dupire formula. We consider this approach in the next Chapter, see also [Itkin (2015)] and references therein. To be practically useful, this construction should guarantee no arbitrage for all strikes and maturities, which is a serious challenge for any model based on interpolation. If the no-arbitrage condition is satisfied, then the LV surface can be calculated using Eq.(3.2) below, which is equivalent to, but more convenient than, the original Dupire formula. The second approach relies on the direct solution of the Dupire equation using either analytical or numerical methods. The advantage of the latter approach is that it guarantees no-arbitrage. However, the problem of the direct solution can be ill-posed, [Coleman *et al.* (2001)], and is rather computationally expensive.

An additional difficulty with both approaches is that the calibration algorithm has to be fast in order to be practically useful. On the one hand, analytical or numerical solutions of the Dupire equation are naturally numerically expensive. On the other hand, building a no-arbitrage IV surface could also be surprisingly numerically challenging, because it requires solving a rather involved constrained optimization problem, see [Itkin (2015)]. An additional complication arises from the fact that in the wings the implied variance surface should be at most linear in the normalized strike, [Lee (2004)].

In this chapter we consider a particular approach where calibration is provided by solving the Dupire equation semi-analytically, while the next one describes a parametric approach. Obviously, as the Dupire equation in continuous in the (K, T) space, it cannot be solved given a discrete set of the market quotes. In other words, the local volatility function $\sigma(K, T)$ should be continuous and known. This problem can be overcome if, when solving the Dupire equation, we make some assumptions about the shape of this function. For instance, a simple approach proposed in [Andreasen and Huge (2011)] assumes that the Dupire equation is first discretized by using an implicit finite-difference scheme. This translates the local volatility function into arbitrage consistent prices for a discrete set of expiries but it does not directly specify the option prices between the expiries. These gaps are filled by further assuming the local volatility function to be piecewise constant in time.

In LS2011 this methodology was improved by assuming the local volatility function to be piecewise constant in the strike space. Then the Dupire equation could be solved analytically. Thus, a continuous local volatility function in the strikes space can be obtained by calibration. This approach was further improved and extended (also performance-wise) in [Itkin and Lipton (2018)]. In what follows, we focus on describing the approach of [Itkin and Lipton (2018)] and their results obtained.

To construct a semi-analytical solution of the Dupire equation we extend the approach proposed in LS2011, which is based on the direct solution of the transformed Dupire equation. In LS2011 a piecewise constant LV surface is chosen, and an efficient semi-analytical method for calibrating this surface to the sparse market data is proposed. However, one can argue that ideally the LV function should be continuous in the log-strike space. Below we demonstrate how to extend LS2011 approach by assuming that the local variance is piecewise linear in the log-strike space, so that the corresponding LV surface is continuous in the strike direction (but not in the time direction). While derivatives of the LV function with respect to strike have discontinuities, the option prices, Deltas and Gammas are continuous. This is to compare with LS2011 where the option prices and deltas are continuous while the option Gammas are discontinuous. We also allow for non-zero interest rates and proportional dividends.

The chapter is organized as follows. Section 3.1 introduces the Dupire equation and discusses a general approach to constructing the LV surface. Section 3.2 considers all necessary steps for solving the Dupire equation. Section 3.3 introduces a no-arbitrage interpolation of the source term, which naturally appears when the Laplace transform in time is used, and shows that using this interpolation all the integrals containing this source term can be obtained in a closed form. Section 3.4 considers a special case when the slope of the local variance on some interval is small, so the linear local variance function on this interval becomes flat. Section 3.5 discusses various asymptotic results which are useful for constructing the general solution of the Dupire equation. Section 3.6 is devoted to the calibration of the model and also describes how to get an educated initial guess for the optimizer. Since computing the inverse Laplace transform could be expensive for small time intervals, Section 3.7 describes an asymptotic solution obtained in this limit in [Gatheral *et al.* (2012)] and shows how to use it for our purposes. Section 3.8 describes numerical results for a particular set of market data.

3.1   Local volatility surface

As a general building block for constructing the local volatility surface we consider Dupire's (forward) equation for the Put option price P which is a function of the strike price K and the time to maturity T, [Dupire (1994)].

We assume that the underlying stock process S_t under the risk neutral measure is governed by the following stochastic differential equation

$$dS_t = (r - q)S_t dt + \sigma(S_t, t)S_t dW_t, \qquad S_0 = S,$$

where $r \geq 0$ is a constant risk free rate, $q \geq 0$ is a constant continuous dividend yield, σ is a given local volatility function, and W_t is the standard Brownian motion. We recall that the Dupire forward equation derived in Section 1.1 for the option Put price $P(K, T)$ reads

$$P_T(K, T) = \left\{ \frac{1}{2}\sigma^2(K, T)K^2 \frac{\partial^2}{\partial K^2} - (r - q)K \frac{\partial}{\partial K} - q \right\} P(K, T), \quad (3.1)$$

$$(K, T) \in (0, \infty) \times [0, \infty),$$

subject to the initial and boundary conditions

$$P(K, 0) = (K - S_0)^+,$$

$$P(0, T) = 0, \quad P(K, T)_{K \uparrow \infty} = KD, \quad D = e^{-rT},$$

where $S_0 = S_t|_{t=0}$, and D is the discount factor.

If the market quotes for $P(K, T)$ are known for all K, T, then the LV function $\sigma(K, T)$ can be uniquely determined everywhere by inverting Eq.(3.1).[2] However, in practice, the known set of market quotes is a discrete set of pairs (K_i, T_j), $i = 1, \ldots, n_j, j = 1, \ldots, M$, where n_j is the total number of known quotes for the maturity T_j, which obviously doesn't cover all K, T. So the form of $\sigma(K, T)$ remains unknown.

In order to address this issue, it is customary to choose a functional form of $\sigma(K, T)$ for the corresponding time slice. For instance, in LS2011 $\sigma(K, T)$ is assumed to be a piecewise constant function of K, T. The authors propose a general methodology of solving Eq.(3.1) for their chosen explicit form of $\sigma(K, T)$ by using the Carson-Laplace transform in time and Green's function method in space. This opens the door for using a version of the least-square method for the calibration routine. Of course, by construction, it makes the whole local volatility surface discontinuous at the boundaries of the tiles, and flat in the wings. While the former feature, in itself, is not necessarily an issue, but should be avoided if possible, the latter feature is somewhat more troubling, since, it is shown in [De Marco et al. (2013); Gerhold and Friz (2015)], that the asymptotic behavior of the local variance is linear in the log strike at both $K \to \infty$ and $K \to 0$. While the result for $K \to 0$ is shown to be true at least for the Heston and Stein-Stein models,

---

[2]If the Call option market prices are given for some strikes and maturities, we can use Call-Put parity in order to convert them to Put prices, since for calibration we usually use vanilla European option prices.

the result for $K \to \infty$ directly follows from Lee's moment formula for the implied variance v_{BS}, [Lee (2004)], and the representation of σ^2 via the total implied variance $w = v_{BS}T$ [Lipton (2001); Gatheral (2006)]

$$w_L \equiv \sigma^2(T,K)T = \frac{T\partial_T w}{\left(1 - \frac{X\partial_X w}{2w}\right)^2 - \frac{(\partial_X w)^2}{4}\left(\frac{1}{w} + \frac{1}{4}\right) + \frac{\partial_X^2 w}{2}}, \quad (3.2)$$

where $w = w(X,T), X = \log K/F$ and $F = Se^{(r-q)T}$ is the stock forward price. Therefore, having a flat local volatility deep in the wings should be avoided if possible.

That is why, here we consider a continuous, piecewise linear local variance $v = \sigma^2(X,T)$ in the spatial variable X for a fixed $T = const$. This allows us to match the asymptotic behavior of v in the wings as well as build the whole surface which is much smoother than in the piecewise constant case. Also, in LS2011 the interest rates and dividends are assumed to be zero, while here we take them into account.

3.2   Solution of Dupire's equation

Introducing a new dependent variable

$$B(X,T) = e^{-X/2}(KD - P(X,T))/Q, \qquad Q = Se^{-qT},$$

which is a scaled covered Put, the problem in Eq.(3.1) can be re-written as follows

$$B_T - \frac{1}{2}vB_{XX} + \frac{1}{8}vB = 0, \quad (3.3)$$

$$B(X,0) = \frac{K - (K - S)^+}{S}e^{-X/2} = e^{-X/2}\mathbf{1}_{X>0} + e^{X/2}\mathbf{1}_{X\le 0},$$

$$B(X,T)_{X\downarrow-\infty} = 0, \quad B(X,T)_{X\uparrow\infty} = 0, \quad (X,T) \in (-\infty,\infty) \times [0,\infty).$$

A similar transformation is used in [Lipton (2002)] in order to solve the backward Black-Scholes equation. Suppose that there are option price quotes (at least for one strike) for M different maturities $T_1,\ldots,T_M$.[3] Also suppose that for each T_j the market quotes are provided at X_i, $i = 1,\ldots,n_j$.[4] Then the corresponding continuous piecewise linear local variance function $v_j(X)$[5] on the interval $[X_i, X_{i+1}]$ reads

$$v_{j,i}(X) = v_{j,i}^0 + v_{j,i}^1 X, \quad (3.4)$$

---

[3]We assume the maturities are sorted in the increasing order.

[4]The strikes also are assumed to be sorted in the increasing order.

[5]Here in the notation we drop off the dependence of v on T since T is given, and hopefully it doesn't bring any confusion.

where we use the super-index 0 to denote a level v^0, and the super-index 1 to denote a slope v^1. Subindex $i = 0$ in $v_{j,0}^0, v_{j,0}^1$ corresponds to the interval $(-\infty, X_1]$. Since $v_j(X)$ is continuous, we have

$$v_{j,i}^0 + v_{j,i}^1 X_{i+1} = v_{j,i+1}^0 + v_{j,i+1}^1 X_{i+1}, \quad i = 0, \ldots, n_j - 1. \tag{3.5}$$

The first derivative of $v_j(X)$ experiences a jump at the points X_i, $i \in \mathbb{Z} \cap [1, n_j]$.

Further, assume that $v(X, T)$ is a piecewise constant function of time, i.e. $v_{j,i}^0, v_{j,i}^1$ don't depend on T on the intervals $[T_j, T_{j+1})$, $j \in [0, M-1]$, and jump to new values at the points T_j, $j \in \mathbb{Z} \cap [1, M]$. In the original independent variables K, T this condition implies that

$$v(K_i, T) \equiv v_{j,i} = v_{j,i}^0 + v_{j,i}^1 \left[\log(K_i/S) - (r-q)T\right], \quad T \in [T_j, T_{j+1}),$$

i.e. that the local variance is a (discontinuous) piecewise linear function of time T. In other words, in the original log-variables $(\log K, T)$ the function $v(\log K, T)$ is piecewise linear in both variables, while in the transformed variables (X, T) the function $v(X, T)$ is piecewise linear in X and piecewise constant in T. Thus, X can be viewed as an automodel variable.[6]

With the above assumptions in mind, Eq.(3.3) can be solved by induction. One starts with $T_0 = 0$, and on each time interval $[T_{j-1}, T_j]$, $j \in \mathbb{Z} \cap [1, m]$ solves the modified problem for $B_j(X, \tau)$

$$B_{j,\tau}(X, \tau) - \frac{1}{2}v_j(X)B_{j,XX}(X, \tau) + \frac{1}{8}v_j(X)B_j(X, \tau) = 0, \tag{3.6}$$

$$B_1(X, 0) = B(X, 0), \quad B_j(X, 0) = B_{j-1}(X, \tau_{j-1}), \ j > 1$$

$$B(X, \tau)_{X \to \pm\infty} = 0, \quad (X, \tau) \in (-\infty, \infty) \times [0, \tau_j],$$

where τ is a continuous time T counted from T_{j-1}, so $\tau_j \equiv T_j - T_{j-1}$, and B_j is the solution of Eq.(3.3) corresponding to the time interval $T_{j-1} \leq T \leq T_j$, $j \in \mathbb{Z} \cap [1, m]$.

To solve Eq.(3.6), similarly to LS2011, we use the Laplace-Carson transform $\hat{B} = \mathcal{L}(p)\{B\}$ of Eq.(3.6) (for application of the Laplace transform to derivatives pricing, see [Lipton (2001)]) to obtain

$$-\frac{1}{2}v_j(X)\hat{B}_{j,XX} + \left(\frac{v_j(X)}{8} + p\right)\hat{B}_j = pB_{j-1}(X, \tau_{j-1}), \tag{3.7}$$

$$\hat{B}(X, p)_{X \uparrow \pm\infty} = 0.$$

Since $v(X)$ is a piecewise linear function, the solution of Eq.(3.7) can also be constructed separately for each interval $[X_{i-1}, X_i]$. By taking into account

---

[6]This terminology is borrowed from aerodynamics and physics of gases and fluids.

the explicit representation of $v(X)$ in Eq.(3.4), from Eq.(3.7) for the i-th spatial interval we obtain

$$(b_2 + a_2 X)\hat{B}_{j,XX} + (b_0 + a_0 X)\hat{B}_j = pB_{j-1}(X, \tau_{j-1}), \qquad (3.8)$$

$$b_2 = -v_{j,i}^0/2, \quad a_2 = -v_{j,i}^1/2, \quad b_0 = p + v_{j,i}^0/8, \quad a_0 = v_{j,i}^1/8.$$

Eq.(3.8) is an *inhomogeneous* Laplace equation, [Polyanin and Zaitsev (2003)], page 155. It is well known that if $y_1 = y_1(X)$, $y_2 = y_2(X)$ are two fundamental solutions of the corresponding *homogeneous* equation, then the general solution of Eq.(3.8) can be represented as

$$\hat{B}_j(p) = C_1 y_1 + C_2 y_2 + p I_{12} \qquad (3.9)$$

$$I_{12} = y_2 \int \frac{y_1 B_{j-1}(X, \tau_{j-1})}{(b_2 + a_2 X)W} dX - y_1 \int \frac{y_2 B_{j-1}(X, \tau_{j-1})}{(b_2 + a_2 X)W} dX,$$

where $W = y_1(y_2)_X - y_2(y_1)_X$ is the so-called Wronskian corresponding to the chosen solutions y_1, y_2. Thus, the problem is reduced to finding suitable fundamental solutions of the homogeneous Laplace equations. Based on [Polyanin and Zaitsev (2003)], if $a_2 \neq 0$ and $a_0 \neq 0$, the general solution reads

$$\hat{B}_j = e^{kX} \mathcal{J}(a, 0, 2k(\mu - X)), \qquad (3.10)$$

$$k = \sqrt{-a_0/a_2} = \pm\frac{1}{2}, \quad \mu = -\frac{b_2}{a_2} = \frac{v_{j,i}^0}{v_{j,i}^1}, \quad a = \frac{b_2 k^2 + b_0}{2a_2 k}.$$

Here $\mathcal{J}(a, b, z)$ is an arbitrary solution of the degenerate hypergeometric equation, i.e., Kummer's function, [Abramowitz and Stegun (1964)], page 504 with a, b being some constants and z being an independent variable. Two types of Kummer's functions are known, namely $M(a, b, z)$ and $U(a, b, z)$, which are Kummer's functions of the first and second kind.[7]

3.2.1   *Numerically satisfactory solutions*

To explicitly represent Eq.(3.9), among all possible fundamental pairs of the solutions given in Eq.(3.10), for every spatial interval we have to determine the pair that is numerically satisfactory, [Olver (1997)]. Since our boundary conditions are set at positive and negative infinity, we need a numerically satisfactory solution for the whole real line. However, it is well known that a single pair of Kummer's functions cannot be numerically satisfactory throughout the whole real line. To overcome this problem, a

---

[7]Due to the linearity of the degenerate hypergeometric equation any linear combination of Kummer's functions also solves this equation.

combined solution can be constructed; below we describe our construction in some detail.

As a preliminary notice, observe that based on the definitions in Eq.(3.10), Eq.(3.8) the variable z can be re-written as $z = -2kv_{j,i}/v^1_{j,i}$. Based on the usual shape of the local variance curve and its positivity, see, e.g., [Itkin (2015)] and references therein, for $X \to -\infty$, we expect that $v^1_{j,i} < 0$. Similarly, for $X \to \infty$ we expect that $v^1_{j,i} > 0$. In between these two infinite limits the local variance curve for a given maturity T_j is assumed to be continuous, but the slope of the curve could be both positive and negative. Also $v_{j,i} \geq 0 \quad \forall X \in \mathbb{R}$, and $a = -p/(kv^1_{j,i})$. With these observations in mind, we now present our methodology.[8]

3.2.1.1 $v^1_{j,i} < 0$

For every interval where $v^1_{j,i} < 0$, i.e. $\forall \ i \in \mathbb{Z} \cap [1, n_j]$ such that $X \in [X_{i-1}, X_i]$, $X_0 = -\infty$, $X_i \leq X_{m_j}$, as the first independent solution of the Kummer equation we take $Y_1(z) = zU(a+1, 2, z)$ with $k = 1/2$, which means $a > 0$.[9] From the definition of z it follows that $X = \mu - z/(2k) = \mu - z$. Thus, when $X \to -\infty$ we have $z \to \infty$ and $e^{kX}Y_1(z) \to 0$.

This solution is numerically stable across the whole interval $X < X_{m_j}$ except the point $z = 0$, which corresponds to $X = \mu$, or, equivalently, $v_{j,i} = 0$; this point belongs to our interval if $\mu < X_{m_j}$.[10] At $z = 0$ the solution has a branch point, [Olver (1997)]. The principal branch of $U(a, b, z)$ corresponds to the principal value of z^{-a} and has a cut in the z-plane along the interval $(-\infty, 0]$. However, one can observe, that at $z = 0$ Eq.(3.8) becomes a degenerated ODE, and its solution immediately reads

$$\hat{B}_j = p \frac{B_{j-1}(X, \tau_{j-1})}{b_0 + a_0 X} = B_{j-1}(X, \tau_{j-1}). \tag{3.11}$$

Therefore, we can exclude this special case from the below consideration, while if this case were to occur during the actual calibration, we just use the special solution given by Eq.(3.11) instead of the general solution.

As the second independent solution of Kummer's equation for $v^1_{j,i} < 0$ (or $X < X_{m_j}$) we have two choices: $Y_2(z) = ze^zU(1-a, 2, -z)$ or $Y_2(z) = zM(a+1, 2, z)$. It can be shown that if we take the former with $k = -1/2$

---

[8]The case where the local volatility is flat on some interval, i.e. $a_2 = 0$, and $a \to \infty$, is considered in Section 3.4.

[9]Since in our case $b = 0$, by Kummer's transformation, [Olver (1997)], page 254, [Abramowitz and Stegun (1964)], page 505, $U(a, 0, z) = zU(a+1, 2, z)$.

[10]As the local variance is linear and non-negative, either this point is at the edge of the interval, or the local variance is flat and vanishes on this interval.

(so $a < 0$ and $X = \mu + z$), then two solutions $e^{-X/2}Y_2(X)$ and $e^{X/2}Y_1(X)$ differ just by a constant e^μ, so that they are not independent. Therefore, we are compelled to keep $k = 1/2$, $a > 0$, and $X = \mu - z$. However, then the function $ze^{z+kX}U(1 - a, 2, -z)$ diverges at both $X \to -\infty$ and at $z \to 0$. Similarly, the function $zM(a + 1, 2, z)$ also diverges at $X \to -\infty$, but is numerically stable at $z \to 0$.

Thus, we have to put $C_2 = 0$ in Eq.(3.9) on the very first interval $(-\infty, X_1]$ to preserve the boundary conditions at $X \to -\infty$. However, the solution $Y_2(z)$ still contributes to I_{12}. In what follows we will use $Y_2(z) = zM(1 + a, 2, z)$, and show that I_{12} converges in the limit $X \to -\infty$.

For future reference, note that the solutions $y_1(z) = e^{kX}zM(a + 1, 2, z)$ and $y_2 = e^{kX}zU(a + 1, 2, z)$ can also be re-written in terms of Whittaker's functions $M_{p,s}(z), W_{p,s}(z)$ [Abramowitz and Stegun (1964)], page 504,

$$y_1(z) = e^{k\mu}M_{-a,1/2}(z), \qquad y_2(z) = e^{k\mu}W_{-a,1/2}(z).$$

3.2.1.2 $v_{j,i}^1 > 0$

For every interval where $v_{j,i}^1 > 0$, i.e. $\forall\, i \in \mathbb{Z} \cap [2, n_j + 1]$ such that $X \in [X_{i-1}, X_i]$, $X_{n_j+1} = \infty$, $X_i > X_{m_j}$, as the first independent solution of Kummer's equation we again take $Y_1(z) = zU(a + 1, 2, z)$ with $k = -1/2$, which means $a > 0$, and $X = \mu + z$. Thus, when $X \to \infty$ we have $z \to \infty$ and $e^{kX}Y_1(z) \to 0$.

Again, this solution is numerically stable on the whole interval $X > X_{m_j}$ except for a singularity at $z = 0$ (if $\mu > X_{m_j}$). However, the solution at $z = 0$ of Eq.(3.8) was already given in the previous subsection.

As far as the second numerically stable solution is concerned, the analysis of the previous subsection is applicable here as well. Therefore, we also take $Y_2(z) = zM(1 + a, 2, z)$ again with $k = -1/2$, so $a > 0$, and $X = \mu + z$. Accordingly, in Eq.(3.9) we put $C_2 = 0$ on the very last interval $[X_{n_j}, \infty)$ to preserve the boundary conditions at $X \to \infty$.

3.2.2 *The combined solution across the whole real line*

The solutions described in the previous section are schematically represented in Table 3.1. Accordingly, for $j > 1$ the solution in Eq.(3.9) on the interval i reads

$$\hat{B}_i = C_{1,i}^{(1)}y_{1,i}(z) + C_{2,i}^{(1)}y_{2,i}(z) + pI_{12,i}^{(1)}(X), \tag{3.12}$$

$$W \equiv W_{1,i} = e^X z^2 W[U(1 + a_i, 2, z), M(1 + a_i, 2, z)] = -\frac{e^{\mu_i}}{\Gamma(a_i + 1)},$$

Table 3.1: Our construction of numerically satisfactory Kummer's pairs. Here "fc" means *from continuity*.

Interval	$v_{j,i}^1$	k	z	y_1	y_2	C_2
$(-\infty, X_1]$	< 0	$1/2$	$\mu - X$	$e^{X/2}zU(a+1,2,z)$	$e^{X/2}zM(a+1,2,z)$	0
$[X_i, X_{i+1}]$	< 0	$1/2$	$\mu - X$	$e^{X/2}zU(a+1,2,z)$	$e^{X/2}zM(a+1,2,z)$	fc
$[X_i, X_{i+1}]$	> 0	$-1/2$	$X - \mu$	$e^{-X/2}zU(a+1,2,z)$	$e^{-X/2}zM(a+1,2,z)$	fc
$[X_{m_j}, \infty)$	> 0	$-1/2$	$X - \mu$	$e^{-X/2}zU(a+1,2,z)$	$e^{-X/2}zM(a+1,2,z)$	0

where $\Gamma(x)$ is the gamma function, $I_{12,i}$ is I_{12} defined in Eq.(3.9) and computed on the interval i, and the superscript$^{(s)}$ in $I_{12,i}^{(s)}$ means that $y_{1,i}(X), y_{2,i}(X)$ (the solutions of the homogeneous equation) in the definition of I_{12} in Eq.(3.9) are taken on the corresponding area (s).[11]

For $j = 1$ the term $B_{j-1}(X, \tau_{j-1})$ should be replaced with $e^{X/2}$ if $X \in [X_{i-1}, X_i], X \le 0$, and with $e^{-X/2}$ if $X \in [X_{i-1}, X_i], X > 0$ in the definition of $I_{12,i}$.

Also in order to satisfy the boundary conditions, see Eq.(3.7), I_{12} in Eq.(3.9) should vanish when $X \to \pm\infty$. This is a subject of the following proposition.

Proposition 3.1. *For $X \to \pm\infty$ the function $I_{12}(X)$, defined in Eq.(3.9), vanishes.*

Proof. First, we intend to prove this Proposition for $j = 1$. In this case the Eq.(3.9) has the form

$$\hat{B} = \begin{cases} C_1 y_1 + C_2 y_2 + p h_1(X), & X \le 0, \\ C_3 y_1 + C_4 y_2 + p h_2(X), & X > 0, \end{cases} \tag{3.13}$$

$$h_i(X) = y_2 I_1(g_i(X)) - y_1 I_2(g_i(X)), \quad i = 1, 2,$$

$$g_1(X) = e^{X/2}, \quad g_2(X) = e^{-X/2},$$

$$I_s(g_l(X)) = \int_\xi^X y_s \frac{g_l(X)}{(b_2 + a_2 X)W} dX, \quad s, l \in \mathbb{Z} \cap [1, 2].$$

Thus, in this case $I_{12}(X) = h_1(X)$ if $X \le 0$, and $I_{12}(X) = h_2(X)$ if $X > 0$. Once this is done, due to the boundary conditions at $X \to -\infty$, the function $\hat{B}(X, \tau_j)$ in Eq.(3.9) tends to $g_1(X)$ in Eq.(3.13), and at $X \to \infty$ we have $\hat{B}(X, \tau_j) \to g_2(X)$. Therefore, at $X \to -\infty$ we have $I_{12}(X) \to h_1(X)$, and at $X \to \infty$, similarly $I_{12}(X) \to h_2(X)$. Thus, the first step of the proof is

---

[11]We use the notation $C_{1,i}^{(l)}, C_{2,i}^{(l)}$ for the integration constants, where super index $l \in \mathbb{Z} \cap [1, 2]$ marks the corresponding area in Fig. 3.1, and the sub index i marks the interval in the X space.

sufficient to prove the Proposition in its entirety. At $X \to -\infty$ (according to Section 3.2.1 this region belongs to the area where $v_{j,i}^1 < 0$) we have $z \to \infty$, and, as follows from Table 3.1 and Eq.(3.13)

$$I_1(g_1(X)) = \int \frac{y_1(X)g_1(X)}{(b_2 + a_2 X)W}dX = \frac{\Gamma(a+1)}{a_2}e^{-\mu/2}\int e^{-z/2}M(1+a,2,z)dz,$$
(3.14)

$$I_2(g_1(X)) = \int \frac{y_2(X)g_1(X)}{(b_2 + a_2 X)W}dX = \frac{\Gamma(a+1)}{a_2}e^{-\mu/2}\int e^{-z/2}U(a+1,2,z)dz.$$

Thus,

$$h_1(z) = \frac{\Gamma(a+1)}{a_2}G(z),$$
(3.15)

$$G(z) \equiv e^{-z/2}M(1+a,2,z)\int e^{-z/2}U(a+1,2,z)dz$$

$$- e^{-z/2}U(1+a,2,z)\int e^{-z/2}M(1+a,2,z)dz.$$

From [Olver (1997)], at $z \to \infty$ we have the following asymptotic series representation

$$U(a,2,z) = \Phi_\infty(z), \qquad \Phi_n(z) \equiv z^{-a}\sum_{s=0}^{n}\frac{(a(a-1))_s}{s!}(-z)^{-s}, \qquad (3.16)$$

$$M(a,2,z) = \Psi_\infty(z), \qquad \Psi_n(z) \equiv \frac{e^z z^{a-2}}{\Gamma(a)}\sum_{s=0}^{n}\frac{(1-a)_s(2-a)_s}{s!}z^{-s},$$

where $(\cdot)_s$ is the Pochhammer symbol.

Let us define the function $G_n(z)$ in the same way as $G(z)$ in Eq.(3.15), but replacing $U(a,2,z) = \Phi_\infty(z)$ with $\Phi_n(z)$. It is clear that $\lim_{n\to\infty} G_n(z) = G(z)$. Substituting Eq.(3.16) into this definition and performing integration term-by-term, we arrive at

$$\int e^{-z/2}\Phi_n(z) = -2^{-a}\sum_{s=0}^{n}f(a,s)2^{n-s}(-1)^s\Gamma(-s-a,z/2), \qquad (3.17)$$

$$\int e^{-z/2}\Psi_n(z) = -(-2)^a\sum_{s=0}^{n}f(a,s)2^{n-s}(-1)^s\Gamma(-s+a,-z/2),$$

$$f(a,s) = \frac{(a(a-1))_s}{s!}.$$

where $\Gamma(a,z)$ is an incomplete gamma function. By [Olver (1997)], at $z \to \infty$ we have

$$\Gamma(a,z) = z^{a-1}e^{-z}\sum_{s=0}^{\infty}(-1)^s\frac{(1-a)_s}{z^s}.$$

Substituting this expression into Eq.(3.17) and collecting terms, we can check that the leading term in this series is $G_n(z) \sim z^{-2}$. Thus, $G_n \to 0$ at $z \to \infty$ as $1/z^2$. Since this convergence rate doesn't depend on n, we can take the limit $n \to \infty$ and see that $G(z) \to 0$ at $z \to \infty$. Since at $k = 1/2$ we have $z = \mu - X$, that means that that $h_1(X) \to 0$ for $X \to -\infty$.

For $h_2(x)$ the representation for $I_1(g_2(X)), I_2(g_2(X))$ is similar to that in Eq.(3.14) and reads

$$I_1(g_2(X)) = \int \frac{y_1(X)g_2(X)}{(b_2 + a_2 X)W} dX = -\frac{\Gamma(a+1)}{a_2} e^{-\mu/2} \int e^{-z/2} U(1+a,2,z)dz,$$

$$I_2(g_2(X)) = \int \frac{y_2(X)g_2(X)}{(b_2 + a_2 X)W} dX = -\frac{\Gamma(a+1)}{a_2} e^{-\mu/2} \int e^{-z/2} M(1+a,2,z)dz.$$

Since we need the limit $z \to \infty$, the convergence of these integrals to zero can be proved similarly to the previous case of $z \to -\infty$. Thus, $h_2(X) \to 0$ for $X \to \infty$. $\qquad\qquad\qquad\qquad\qquad\qquad\qquad\qquad\qquad\qquad\qquad\qquad\qquad$ □

Using these results, we can now proceed to constructing the solution of Eq.(3.7) on the whole real line $X \in [-\infty, \infty]$ by matching solutions on all the intervals.

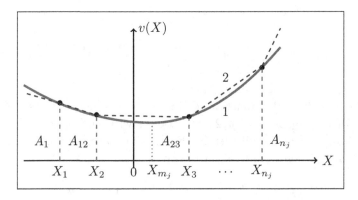

Figure 3.1: Construction of the whole solution of the Dupire equation. 1 (solid curve — the real (unknown) local variance curve, 2 (dashed curve) — a piecewise linear solution.

Suppose that Put prices for $T = T_j$ are known for n_j ordered strikes. Also, first suppose that these quotes are available for both $K > F$ and $K < F$. The location of these strikes on the X line is schematically depicted in Fig. 3.1.

According to the analysis of Section 3.2, on the open interval A_1 the solution of Eq.(3.7) is given by the first line of Table 3.1. It contains one unknown constant $C_{1,1}^{(1)}$ since we put $C_{2,i}^{(1)} = 0$ due to the boundary conditions. The solutions from line 2 in Table 3.1 should be used for all other intervals $A_{k-1,k}$ such that $k \geq 1$, $v_{j,k}^1 \leq 0$. These solutions have two yet unknown constants $C_{1,k}^{(1)}, C_{1,k}^{(2)}$, since X is finite on the corresponding interval, and therefore, both solutions $y_1(X), y_2(X)$ are well-behaved. For X_k, where $v_{j,k}^1 > 0$ and $k \leq m_j$ we use the solution given in the third line of Table 3.1, which also has two yet unknown constants $C_{1,k}^{(2)}, C_{2,k}^{(2)}$ for each interval. Finally, for the interval $X \in [X_{n_j}, \infty))$, we use the solution in the last line of Table 3.1. Again, to obey the boundary conditions we must set $C_{2,n_j+1}^{(2)} = 0$.

Thus, we have $2n_j$ unknown constants to be determined. Since the local volatility function v_i is continuous at the points X_i, $i = 1, \ldots, n_j$, so should be the solution $\hat{B}(X, \tau_j)$. Therefore, we require that at the points z_i, $i = 1, \ldots, n_j$ the solution and its first derivative in X should be a continuous function of X. Thus, the above constants solve a system of $2n_j$ algebraic equations. This system has a special structure that allows one to reduce its LHS matrix to the upper triangular form (actually even the upper banded form). Therefore, it can be efficiently solved with the linear complexity $O(2n_j)$. For more details, see LS2011.

When computing the first derivatives, we take into account that

$$h_{i,X} = y_{1,X} I_1(g_i(X)) - y_{2,X} I_2(g_i(X)), \quad i = 1, 2,$$

and according to [Abramowitz and Stegun (1964)], page 507

$$\frac{\partial M(a, b, z)}{\partial z} = \frac{a}{b} M(a+1, b+1, z), \quad \frac{\partial U(a, b, z)}{\partial z} = -aU(a+1, b+1, z).$$

Also, in some special cases which are discussed in the following sections, the solution can be represented in terms of the modified Bessel functions. But it is known, [Abramowitz and Stegun (1964)], that the derivatives of the modified Bessel functions are expressed in closed form via the same set of functions. Therefore, computing the derivatives of the solution doesn't cause any new technical problems.

Note, that in the definition of the integrals I_{12} in Eq.(3.9), for the sake of convenience, we define the low limit of integration $\xi(X)$ as follows. For the interval A_1 we take $\xi = -\infty$. Then for each integral $I_{12}(X_i)$, $i = 2, \ldots, n_j$ we use $\xi_i = X_{i-1}$ (or z_{i-1} if the integral is expressed in z variables, see Appendix). This choice is inspired by the fact that all the parameters $a_i, a_{2,i}, b_{2,i}, \mu_i$ in Eq.(3.9) are constant on the interval $[X_{i-1}, X_i]$.

Also repeat that, for the sake of simplicity, in the above construction we assumed that market quotes are available for a set of strikes with $K < F$, as well as for a set of strikes with $K > F$. However, it could happen that the market provides just a set of strikes such that all $X_i > 0$ or $X_i < 0$. In this case we can construct the whole solution as follows.

Suppose $X_i < X_{m_j}$, $\forall\, i \in \mathbb{Z} \cap [1, n_j]$. Introduce an additional auxiliary point $X_* > X_{m_j}$. Of course, since this is an auxiliary point, the corresponding market quote is not available. Therefore, we don't need to calibrate the local variance at this point. However, introduction of such a point helps to construct the solution on the whole real line, similarly to how it was done above. An unknown constant $C_{1,*}^{(2)}$ again can be found assuming the continuity of the solution at the point X_*, while $C_{2,*}^{(2)}$ should be set to 0 to preserve the boundary conditions. Thus, this trick just helps to construct a numerically stable solution across the region $(-\infty, X_1]$ with $X_1 > X_{m_j}$ (when there are no points $X_i < X_{m_j}$), or across the region $[X_{n_j}, \infty)$ with $X_{n_j} < X_{m_j}$ (when there are no points $X_i > X_{m_j}$).

According to our construction, the options values as well as option deltas and gammas are continuous in X (and, therefore, in S). Indeed, in the above we required $\hat{B}(X, \tau_j), \hat{B}_X(X, \tau_j)$ and v_j to be continuous functions of X. Then, based on Eq.(3.7), $\hat{B}_{XX}$ is also a continuous function of X. Applying the inverse Laplace transform, we obtain that B_{XX} is also continuous in X, and, hence, by the definition of X, in S. Therefore, by the definition of B, P, $\frac{\partial P}{\partial S}$, $\frac{\partial^2 P}{\partial S^2}$ are also continuous. This result demonstrates the additional advantage of our model as compared, e.g., with LS2011, where the options gammas are discontinuous due to only a piecewise continuity of v_j.

3.3 Analytical representation of the integrals $I_{12}(X)$

To compute the RHS term $h(X) = pI_{12}(X)$ at some time step j we need a function $B_{j-1}(X, \tau_{j-1})$ obtained at the previous time step. However, market quotes at T_j and T_{j-1} could be given at different sets of X even if the strikes are same since, by definition, $X = \log(K/F(T))$. Therefore, when computing $pI_{12}(X)$ in Eq.(3.9) by using a numerical quadrature, we need to know the values of $B_{j-1}(X, \tau_{j-1})$ at points X where they have not been calculated yet. There are at least two possible approaches to addressing this issue.

The first approach relies on the fact that the solution $\hat{B}_{j-1}$ is already known for each space interval $[X_{i-1}, X_i]$. Therefore, to compute $B_{j-1}(X)$,

$X_{i-1} < X < X_i$ we can use the inverse Laplace transform method as described below. Also this would require computation of $I_{12}(X, \tau_{j-1})$ since this is a part of the solution for $\hat{B}_{j-1}$. Thus, this method, despite being exact, is very computationally expensive as it requires the inverse Laplace transform and numerical integration embedded into another inverse Laplace transform and numerical integration.

The second approach, which is advocated by the author, uses interpolation to compute $B_{j-1}(X)$ given the values of $B_{j-1}(\bar{X})$, where $\bar{X} = X(T_{j-1})$, and $X = X(T_j)$. In general, linear interpolation would be sufficient. However, as this is shown in Section 2.1, it gives rise to the violation of the no-arbitrage conditions. Therefore, we proceed with a no-arbitrage version which was proposed in [Itkin and Lipton (2018)], and also described in Section 2.1.

Using the definition of X and $B(X,T)$ and some algebra, the interpolation formula in Eq.(2.2) for $B(X)$ can be re-written as

$$B[X, \tau] = \alpha_1^- e^{-X/2} + (\beta_1^+ X + \beta_2^+) e^{X/2}, \tag{3.18}$$

$$\alpha_1^- = \frac{e^{\frac{X_1 + X_3}{2}}}{R} \left[e^{\frac{X_1 + X_3}{2}} (k_3 - k_1) + e^{\frac{X_1}{2}} k_1 B(X_3, T) - e^{\frac{X_3}{2}} k_3 B(X_1, T) \right],$$

$$\beta_1^+ = \frac{1}{R} \left[e^{X_3} - e^{X_1} - e^{\frac{X_3}{2}} B(X_3, T) + e^{\frac{X_1}{2}} B(X_1, T) \right],$$

$$\beta_2^+ = \frac{1}{R} \left[\left(e^{\frac{X_1}{2}} B(X_1, T) - e^{\frac{X_3}{2}} B(X_3, T) \right) \log F + e^{X_1} X_1 - e^{X_3} X_3 \right],$$

$$R = k_1 e^{X_1} - k_3 e^{X_3}.$$

where $k_i = X_i + \log F$. In Fig. 3.2 the relative difference for linear B_L and non-linear B_N interpolations vs. the exact Black-Scholes value B_E is shown as a function of X. It can be seen that in this test the difference is around 3 bps.

Now the expression given by Eq.(3.18) can be substituted into the definition of $I_{12}(X)$ in Eq.(3.9). It turns out that then the corresponding integral can be computed in closed form as we show in the next Section.

3.3.1 Closed form representation of all integrals in $I_{12}(X)$

Here we derive an analytical expression for $I_{12}(X)$ in Eq.(3.9), which takes into account our approximation of $B(X, \tau_{j-1})$ presented in Section 3.3, and

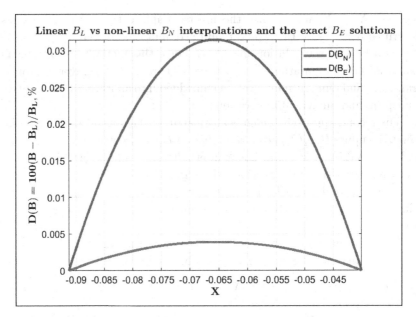

Figure 3.2: Absolute differences $D(B) = B - B_L$ for no-arbitrage non-linear interpolation B_N, and the exact Black-Scholes Put prices B_E, with the linear interpolation B_L. The line $D(B_L) = 0$ corresponds to B_L.

reads

$$I_{12}(X) = y_2 I_1(X) - y_1 I_2(X),$$

$$I_1(X) = \int_\xi^X \frac{y_1 B_{j-1}(X, \tau_{j-1})}{(b_2 + a_2 X)W} dX$$

$$= \int_\xi^X \frac{y_1[\alpha_1^- e^{-X/2} + (\beta_1^+ X + \beta_2^+)e^{X/2}]}{(b_2 + a_2 X)W} dX,$$

$$I_2(X) = \int_\xi^X \frac{y_2 B_{j-1}(X, \tau_{j-1})}{(b_2 + a_2 X)W} dX$$

$$= \int_\xi^X \frac{y_2[\alpha_1^- e^{-X/2} + (\beta_1^+ X + \beta_2^+)e^{X/2}]}{(b_2 + a_2 X)W} dX.$$

Suppose we compute these integrals on the interval $[X_i, X_{i+1}]$, i.e. $X \in [X_i, X_{i+1}]$. As the lower limit of integration ξ it is convenient to choose $\xi = X_i$. Then the coefficient a_2, b_2 are constant on this interval, and so are a, α, β. The homogeneous solutions y_1, y_2 should be chosen according to the analysis of Section 3.2.

$v_{j,i}^1 < 0$ According to Table 3.1, for negative $v_{j,i}^1$ we have

$$y_1 = ze^{X/2}U(a+1, 2, z), \qquad y_2 = ze^{X/2}M(1+a, 2, z),$$

$$W = -\frac{e^\mu}{\Gamma(a_i + 1)}, \qquad z = \mu - X.$$

Therefore,

$$I_2 = -\Gamma(a+1)e^{-\mu} \int \frac{e^{X/2}zM(1+a, 2, z)}{b_2 + a_2 X}[\alpha_1^- e^{-X/2} + (\beta_1^+ X + \beta_2^+)e^{X/2}]dX$$

$$= \frac{\Gamma(a+1)}{a_2}[\alpha_1^- J_0 + \beta_2^+ e^\mu J_1 + e^\mu \beta_1^+ J_2],$$

$$J_0 = \int M(1+a, 2, z)dz, \quad J_1 = \int e^{-z}M(1+a, 2, z)dz,$$

$$J_2 = \int (\mu - z)e^{-z}M(1+a, 2, z)dz = \mu J_1 - J_3,$$

$$J_3 = \int ze^{-z}M(1+a, 2, z)dz.$$

From [Ng and Geller (1970)] after some transformations we obtain

$$J_1 = \int e^{-z}M(1+a, 2, z)dz = \frac{1}{a}e^{-z}M(1+a, 1, z),$$

$$J_0 = \int M(1+a, 2, z)dz = \frac{1}{a}M(a, 1, z),$$

$$J_3 = \int ze^{-z}M(1+a, 2, z)dz = \frac{1}{2}z^2 e^{-z}M(a+2, 3, z).$$

Similarly,

$$I_1 = -\Gamma(a+1)e^{-\mu} \int \frac{e^{X/2}zU(1+a, 2, z)}{b_2 + a_2 X}[\alpha_1^- e^{-X/2} + (\beta_1^+ X + \beta_2^+)e^{X/2}]dX$$

$$= \frac{\Gamma(a+1)}{a_2}e^{-\mu}[\alpha_1^- \mathcal{J}_0 + \beta_2^+ e^\mu \mathcal{J}_1 + \beta_1^+ e^\mu \mathcal{J}_2],$$

$$\mathcal{J}_0 = \int U(1+a, 2, z)dz, \qquad \mathcal{J}_1 = \int e^{-z}U(1+a, 2, z)dz,$$

$$\mathcal{J}_2 = \int Xe^{-z}U(1+a, 2, z)dz = \mu \mathcal{J}_1 - \mathcal{J}_3, \quad \mathcal{J}_3 = \int ze^{-z}U(1+a, 2, z)dz.$$

Again, from [Ng and Geller (1970)] we can obtain

$$\mathcal{J}_0 = \int U(a+1, 2, z)dz = -\frac{1}{a}U(a, 1, z),$$

$$\mathcal{J}_1 = \int e^{-z}U(a+1, 2, z)dz = -e^{-z}U(a, 1, z),$$

$$\mathcal{J}_3 = \int ze^{-z}U(a, 2, z)dz = -z^2 e^{-z}U(a+2, 3, z).$$

$v_{j,i}^1 > 0$ According to Table 3.1, for positive $v_{j,i}^1$ we have

$$y_1 = ze^{-X/2}U(a+1,2,z), \qquad y_2 = ze^{-X/2}M(1+a,2,z),$$

$$W = -\frac{e^\mu}{\Gamma(a+1)}, \qquad z = \mu + X.$$

Hence

$$I_2 = -\Gamma(a+1)e^{-\mu} \int \frac{ze^{-X/2}M(1+a,2,z) \\ \times[\alpha_1^- e^{-X/2} + (\beta_1^+ X + \beta_2^+)e^{X/2}]}{b_2 + a_2 X} dX$$

$$= -\frac{\Gamma(a+1)}{a_2}e^{-\mu}\left[\alpha_1^- e^\mu \mathcal{I}_0 + \beta_2^+ \mathcal{I}_1 + \beta_1^+ \mathcal{I}_2\right],$$

$$\mathcal{I}_0 = \int e^{-z}M(1+a,2,z)dz = J_1, \quad \mathcal{I}_1 = \int M(1+a,2,z)dz = J_0,$$

$$\mathcal{I}_2 = \int (z-\mu)M(1+a,2,z)dz = \mathcal{I}_3 - \mu J_0,$$

$$\mathcal{I}_3 = \int zM(1+a,2,z)dz = \frac{z-1}{a}M(a,1,z) + \frac{1}{a}M(a-1,1,z).$$

Similarly,

$$I_1 = -\Gamma(a+1)e^{-\mu} \int \frac{ze^{-X/2}U(1+a,2,z) \\ \times[\alpha_1^- e^{-X/2} + (\beta_1^+ X + \beta_2^+)e^{X/2}]}{b_2 + a_2 X} dX$$

$$= -\frac{\Gamma(a+1)}{a_2}e^{-\mu}\left[e^\mu \alpha_1^- \mathcal{P}_2 + \beta_2^+ \mathcal{P}_0 + \beta_1^+(\mathcal{P}_3 - \mu \mathcal{P}_0)\right],$$

$$\mathcal{P}_0 = \int U(1+a,2,z)dz = \mathcal{J}_0, \qquad \mathcal{P}_2 = \int e^{-z}U(1+a,2,z)dz = \mathcal{J}_1,$$

$$\mathcal{P}_3 = \int zU(1+a,2,z)dz = -\frac{z}{a}\left(U(a,1,z) + \frac{1}{a-1}U(a,2,z)\right).$$

Sometimes, it could happen that for the new maturity deep out-of-the-money (OTM) or in-the-money (ITM) strikes are positioned outside of the region covered by strikes given for the previous maturity. That means, that no-arbitrage interpolation cannot be used in such a case, while using Eq.(3.18) for extrapolation will lead to arbitrage. This issue can be addressed as follows.

Suppose that at T_j the last strike with a known market quote is K_{j,n_j}. Suppose that at $T_{j+1} > T_j$ we are given a set of new strikes $K_{j+1,1}, ..., K_{j+1,n_{j+1}}$, such that $X_{j+1,l} > X_{j,n_j}, \forall l : n_{j+1} \leq l \leq i$, where i is some integer $i \in \mathbb{Z} \cap [1, n_{j+1}]$. Now introduce an auxiliary point $X_{j,*}$, such that $X_{j,*} > X_{j+1,n_j+1}$ and $X_{j,*} < \infty$. Based on the boundary conditions we can assume that $B(X_*, T_j) = 0$. Then, having this extra auxiliary point, the problem of extrapolation reduces to interpolation which was discussed above. A similar approach can be used at the opposite end when $X_{j+1,l} < X_{j,1}, \ \forall \, l \in \mathbb{Z} \cap [1,i]$ for some $i > 0$. Then the auxiliary point $X_{j,*}$ should be inserted on the interval $-\infty < X_{j,*} < X_{j+1,1}$.

Solution for the first term T_1: For the first term T_1 we don't need interpolation since we know the solution $B(X,0)$ along the whole real line $X \in (-\infty, \infty)$. It is given by the terminal condition in Eq.(3.3) and fits into our interpolation formula in Eq.(3.18) if we set $\alpha_1^- = p\mathbf{1}_{X>0}$ and $\beta_1^+ = 0, \beta_2^+ = p\mathbf{1}_{X\leq 0}$. Thus, in this case the analytical solution for $I_{12}(X)$ is still available.

3.4 Small $|v_{j,i}^1|$

When calibrating the model to the market data, it could happen that some values of $v_{j,i}^1$ become small, so that $|v_{j,i}^1 X_i| \ll 1$.[12] In this case, the solutions considered in Section 3.2 are no longer valid. Therefore, we need to consider Eq.(3.8) which can be represented in the form

$$(1 + \epsilon) \hat{B}_{j,XX} + \left(\kappa^2 - \frac{\epsilon}{4}\right) \hat{B}_j = \frac{p}{b_2} B_{j-1}(X, \tau_{j-1}), \qquad (3.19)$$

where $\kappa = \sqrt{b_0/b_2}$, and for each interval $[X_{i-1}, X_i], \ i \in \mathbb{Z} \cap [2, n_j]$ the parameter ϵ is defined as

$$\epsilon = v_{j,i}^1 X_i / v_{j,i}^0.$$

If $|\epsilon_i| \ll 1$, a general solution of Eq.(3.19) $\hat{B}_j$ can be represented as a power series in ϵ, i.e.,

$$\hat{B}_j = \sum_{s=0}^{\infty} \hat{B}_j^{(s)}(X)\epsilon^s.$$

Zeroth-order approximation: In zeroth-order approximation Eq.(3.19) can be written as

$$\hat{B}_{j,XX}^{(0)} + \kappa^2 \hat{B}_j^{(0)} = \frac{p}{b_2} B_{j-1}(X, \tau_{j-1}),$$

---

[12]The case $v_{j,i}^1 = 0$ is considered in LS2011, where the integrals $I_{12}(X)$ were computed numerically.

So that the corresponding variance is piecewise constant. A general solution of this equation has the form

$$\hat{B}_j^{(0)} = C_1 y_1(X) + C_2 y_2(X) + \frac{p}{b_2} I_{12}(X), \qquad (3.20)$$

$$y_1 = e^{\imath \kappa X}, \quad y_2 = e^{-\imath \kappa X},$$

$$I_{12} = y_2 \int \frac{y_1 B_{j-1}(X, \tau_{j-1})}{W} dX - y_1 \int \frac{y_2 B_{j-1}(X, \tau_{j-1})}{W} dX.$$

Obviously, for these y_1, y_2 (which are always numerically satisfactory), we have $W[y_1, y_2] = -2\imath \kappa$. Again, we use the no-arbitrage interpolation of the solution obtained at the previous time step to compute $I_{12}(X)$ explicitly:

$$I_1(X, \kappa) = \int e^{\imath \kappa X} B_{j-1}(X, \tau_{j-1}) \frac{dX}{W}$$

$$= -\frac{1}{2\imath \kappa} \int e^{\imath \kappa X} [\alpha_1^- e^{-X/2} + (\beta_1^+ X + \beta_2^+) e^{X/2}] dX$$

$$= -\frac{1}{2\imath \kappa} \left[\frac{\alpha_1^-}{\delta_-} e^{\delta_- X} + \frac{\beta_1^+ (\delta_+ X - 1) + \delta_+ \beta_2^+}{\delta_+^2} e^{\delta_+ X} \right], \quad \delta_\pm = \imath \kappa \pm 1/2,$$

$$I_2(X) = -\frac{1}{2\imath \kappa} \int e^{-\imath \kappa X} [\alpha_1^- e^{-X/2} + (\beta_1^+ X + \beta_2^+) e^{X/2}] dX$$

$$= \frac{1}{2\imath \kappa} \left[\frac{\alpha_1^-}{\delta_+} e^{-\delta_+ X} + \frac{\beta_1^+ (\delta_- X + 1) + \delta_- \beta_2^+}{\delta_-^2} e^{-\delta_- X} \right],$$

$$I_{12}(X) = e^{-\imath \kappa X} I_1(X, \kappa) - e^{\imath \kappa X} I_2(X, \kappa) = -\frac{1}{\imath \kappa} \left[A_- e^{-X/2} + A_+ e^{X/2} \right],$$

$$A_- = \alpha_1^- \left(\frac{1}{\delta_+} + \frac{1}{\delta_-} \right),$$

$$A_+ = \frac{\beta_1^+ (\delta_- X + 1) + \delta_- \beta_2^+}{\delta_-^2} + \frac{\beta_1^+ (\delta_+ X - 1) + \delta_+ \beta_2^+}{\delta_+^2}.$$

These solutions can be considered as a further improvement of LS2011, since (i) they embed a no-arbitrage interpolation, and (ii) this interpolation allows computation of the source terms in closed form. Obviously, performance-wise such an approach significantly speeds up the calculations.

First-order approximation: In first-order approximation in $\epsilon \ll 1$, Eq.(3.19) has the form

$$X \hat{B}_{j,XX}^{(1)} + (2 + \kappa^2 X) \hat{B}_j^{(1)} = X \mathcal{B}^{(0)}(X),$$

$$\mathcal{B}^{(0)}(X) = \left(\kappa^2 + \frac{1}{4} \right) \hat{B}_j^{(0)} - \frac{p}{b_2} B_{j-1}(X, \tau_{j-1}).$$

If $|X| \ll 1$, then

$$\hat{B}_j^{(1)} = \frac{X}{2} \mathcal{B}^{(0)}(X).$$

Otherwise, the solution to this equation reads, see [Polyanin and Zaitsev (2003)], page 155

$$\hat{B}_j^{(1)} = C_2 + C_1 y_1(X) + I_{12}(X), \tag{3.21}$$

$$y_1 = -\kappa^2 \mathrm{Ei}(-\kappa^2 X) - \frac{e^{-\kappa^2 X}}{X}, \qquad W = \frac{e^{-\kappa^2 X}}{X^2},$$

$$I_{12} = \int \frac{y_1 X \mathcal{B}^0(X)}{W} dX - y_1 \int \frac{X \mathcal{B}^0(X)}{W} dX,$$

where $\mathrm{Ei}(X)$ is the exponential integral, [Abramowitz and Stegun (1964)], page 227. If $\kappa^2 > 0$ then C_1 should be set to zero when $X \to -\infty$, i.e., on the very first interval. If $\kappa^2 < 0$ C_1 should be set to zero when $X \to \infty$, i.e., on the very last interval.

3.5 Large values of the parameter $|a|$

In many practical situations the parameter $|a|$ in Eq.(3.10) can become large. Indeed, it follows from the analysis of Section 3.2.2 that $|a_i| = 2p/|v_{j,i}^1|$. The values of p we are interested in can be estimated by taking into account the fact that for computation of the inverse Laplace transform we use the Gaver-Stehfest algorithm described in Section 3.6. Then, by virtue of Eq.(3.28), $p = s(\log 2)/\tau_j$, where s runs from 1 to $N = 12$. Therefore, for a typical value of $\tau_j = 0.1$, p changes in the range from 7 to 83. At the same time, usually $|v_{j,i}^1| = O(0.1)$, so that $|a| \gg 1$.

From [Abramowitz and Stegun (1964); Olver (1997)] we know that for large values of a the value of $U(a+1, 2, z)$ is very small, while the value of $U(1-a, 2, -z)$ is very big. Therefore, the computation of unknown constants $C_{1,i}^{(2)}, C_{2,i}^{(2)}$ is difficult, because (i) it requires high-precision arithmetic, and (ii) it is pretty unstable. On the other hand, in this case we have a small parameter $1/|a| \ll 1$ in Eq.(3.8), so we can find an asymptotic solution of Eq.(3.8).

We start with a rigorous definition of the small parameter $\varepsilon \equiv -k v_{j,i}^1/p$. Here, when choosing the sign of k, we should not rely on the analysis of Section 3.2.2, because we only require convergence of our asymptotic solution when $\varepsilon \to 0$. Below we assume that $|\varepsilon| \ll 1$.[13] With this definition in

---

[13]In what follows, for simplicity we omit the modulo, i.e. by saying that ε is small we mean that $|\varepsilon| \ll 1$.

mind, and using definitions in Eq.(3.10), we re-write Eq.(3.8) in the form

$$\epsilon \bar{X} \hat{B}_{j,XX} + \left(2k - \frac{1}{4}\epsilon \bar{X}\right) \hat{B}_j = 2k B_{j-1}(X, \tau_{j-1}), \qquad \bar{X} = X - \mu, \quad (3.22)$$

where $|2k| = 1$. This equation belongs to the class of singularly perturbed differential equations, [Wasow (1987)]. It can be solved by using either the method of matching asymptotic expansions, [Nayfeh (2000)], or the method of boundary functions, [Vasil'eva *et al.* (1995)] which we will use below.

The need for a special method is due to the fact that for a regular asymptotic expansion of the unknown function $\hat{B}(X, \tau)$ in a series in powers of the small parameter ε, zeroth-order approximation yields $\hat{B}^{(0)}(X, \tau_j) = B_{j-1}(X, \tau_{j-1})$. Here the superscript $^{(0)}$ denotes the order of the approximation. Obviously, this solution, which doesn't does not depend on any free parameter, is incorrect in the vicinity of the end points of the interval $[X_{i-1}, X_i]$, where the solution and its first derivative have to be continuous functions of X. So we don't have any degrees of freedom to satisfy this continuity. That is why Eq.(3.22) belongs to the class of singularly perturbed differential equations, which cannot be solved by using regular expansions in powers of ε.

Following [Vasil'eva *et al.* (1995)], we represent the solution of Eq.(3.22) on the interval $[X_{i-1}, X_i]$ in the form

$$\hat{B}(X, \tau_j) = \sum_{s=0}^{\infty} \varepsilon^s \hat{B}^{*,s}(X, \tau_j) + \sum_{s=0}^{\infty} \varepsilon^s \Pi^{(s)}(x_{i-1}, \tau_j) + \sum_{s=0}^{\infty} \varepsilon^s \Xi^{(s)}(x_i, \tau_j).$$
$$(3.23)$$

Here $\hat{B}^*(X, \tau_j)$ is the solution of the so-called "reduced" equation, while $\Pi(x_{i-1}, \tau_j)$ and $\Xi(x_i, \tau_j)$ are the so-called boundary functions. The boundary functions vanish far away from the boundaries X_{i-1}, X_i when $\varepsilon \to 0$. On the other hand, they are needed to ensure that the solution satisfies the boundary conditions. For any small fixed $\varepsilon \ll 1$, $\varepsilon \neq 0$ the asymptotic solution is an approximation of the exact solution which can be obtained up to $O(\varepsilon^N)$ with N being an arbitrary positive integer, see [Vasil'eva *et al.* (1995)].

Also in Eq.(3.23) $x_{i-1} = (X - X_{i-1})/\sqrt{\varepsilon}$ is the stretched distance to the left boundary, and $x_i = (X - X_i)/\sqrt{\varepsilon}$ is the stretched distance to the right boundary.

Based on the method of [Vasil'eva *et al.* (1995)], in zeroth-order approximation the reduced equation, which follows from Eq.(3.22) at $\varepsilon \to 0$, has a trivial solution $\hat{B}^{,0*}(X, \tau_j) = B_{j-1}(X, \tau_{j-1})$. Then, from Eq.(3.22) the boundary function $\Pi^{(0)}(x, \tau_j)$ solves the equation

$$(x - \mu_i)\Pi_{xx}^{(0)}(x, \tau_j) + 2k\Pi^{(0)}(x, \tau_j) = 0. \tag{3.24}$$

The latter has the following solution, [Polyanin and Zaitsev (2003)]

$$\Pi^{(0)}(x, \tau_j) = C_1\phi_{i-1}(x)I_1(2\phi_{i-1}(x)) + C_2\phi_{i-1}(x)K_1(2\phi_{i-1}(x)), \tag{3.25}$$

$$\phi_{i-1}^2(x) \equiv -2k\,(x - \mu_i) = -\frac{2k}{\sqrt{\varepsilon}}\left(X - X_{i-1} - \sqrt{\varepsilon}\mu_i\right).$$

Here C_1, C_2 are two integration constants, and $I_1(x), K_1(x)$ are the modified Bessel functions of the first and second kind.

We must prove that $\Pi^{(0)}(x, \tau_j) \to 0$ when $\varepsilon \to 0$. Based on [Abramowitz and Stegun (1964)], page 358, we know that this is true for $K_1(2\phi_{i-1}(x))$ if $k < 0$ since $X > X_{i-1}$, but not for $I_1(2\phi_{i-1}(x))$. Therefore, in Eq.(3.25) we must put $C_1 = 0$, and $k = -1/2$. Note, that for $v_{j,i}^1 < 0$ $\varepsilon \in \mathbb{C}$, but this is not a problem.

Similar arguments show that for $\Xi^{(0)}(x, \tau_j)$ in zeroth-order approximation in ε the solution reads

$$\Xi^{(0)}(x, \tau_j) = C_2\phi_i(x)K_1(2\phi_i(x)),$$

$$\phi_i^2(x) = -\frac{1}{\sqrt{\varepsilon}}\left(X - X_i - \sqrt{\varepsilon}\mu_i\right).$$

Thus, finally, zeroth-order asymptotic solution of Eq.(3.22) has the form

$$\hat{B}^{(0)}(X, \tau_j) = B_{j-1}(X, \tau_{j-1}) + C_1\phi_{i-1}(x)K_1(2\phi_{i-1}(x))$$
$$+ C_2\phi_i(x)K_1(2\phi_i(x)).$$

The unknown constants C_1, C_2 can be found using the method described in the next section. The values of $B_{j-1}(X, \tau_{j-1})$ at the points X_{i-1}, X_i can be obtained by using our no-arbitrage interpolation described in Section 3.3.

The next approximations in ε can also be constructed based on the general method of [Vasil'eva *et al.* (1995)]. The reduced equation now reads

$$\bar{X}\hat{B}_{XX}^{(*,0)}(X, \tau_j) - \frac{1}{4}\bar{X}\hat{B}^{(*,0)}(X, \tau_j) + 2k\hat{B}^{(*,1)}(X, \tau_j) = 0,$$

with the obvious solution

$$\hat{B}^{(*,1)}(X, \tau_j) = \frac{1}{2k}\bar{X}\left[\frac{1}{4}\partial_{X,X}\hat{B}_{j-1}(X, \tau_{j-1}) - \hat{B}_{j-1}(X, \tau_{j-1})\right].$$

As $\hat{B}_{j-1}(X, \tau_{j-1})$ solves Eq.(3.7), we can represent $\partial_{X,X}\hat{B}_{j-1}(X, \tau_{j-1})$ with $j > 1$ in the form

$$\partial_{X,X}\hat{B}_{j-1}(X, \tau_{j-1}) = -\frac{2p}{v_{j-1,i}(X)}B_{j-2}(X, \tau_{j-2}) - \left(\frac{2p}{v_{j-1,i}(X)} + \frac{1}{4}\right)\hat{B}_{j-1}.$$

The equation for $\Pi^{(1)}(x, \tau_j)$ is

$$(x - \mu_i)\Pi_{xx}^{(1)}(x, \tau_j) + 2k\Pi^{(1)}(x, \tau_j) = \frac{1}{4}(x - \mu_i)\Pi_{xx}^{(0)}(x, \tau_j) = -\frac{k}{2}\Pi^{(0)}(x, \tau_j).$$

This equation is similar to Eq.(3.24), the only difference being that now it is inhomogeneous. Accordingly, its solution reads

$$\Pi^{(1)}(x, \tau_j) = \phi_{i-1}(x)K_1(2\phi_{i-1}(x)) + I_{i-1}^{(1)}, \tag{3.26}$$

$$I_{i-1}^{(1)} = -\frac{k}{2}\left\{ \phi_{i-1}(x)I_1(2\phi_{i-1}(x)) \int \frac{K_1(2\phi_{i-1}(x))}{\phi_{i-1}(x)}\Pi^{(0)}(x, \tau_j)dx \right.$$

$$\left. - \phi_{i-1}(x)K_1(2\phi_{i-1}(x)) \int \frac{I_1(2\phi_{i-1}(x))}{\phi_{i-1}(x)}\Pi^{(0)}(x, \tau_j)dx \right\}$$

$$= -\frac{k}{2}\left\{ \phi_{i-1}(x)I_1(2\phi_{i-1}(x)) \int K_1^2(2\phi_{i-1}(x))dx \right.$$

$$\left. - \phi_{i-1}(x)K_1(2\phi_{i-1}(x)) \int I_1(2\phi_{i-1}(x))K_1(2\phi_{i-1}(x))dx \right\}.$$

From [Prudnikov *et al.* (1986)] we have

$$\int K_1^2(2\phi_{i-1}(x))dx = \phi_{i-1}(x)K_1^2(2\phi_{i-1}(x)) - K_0(2\phi_{i-1}(x))K_2(2\phi_{i-1}(x)),$$

$$\int I_1(2\phi_{i-1}(x))K_1(2\phi_{i-1}(x))dx = \int I_1(2\sqrt{x - \mu_i})K_1(2\sqrt{x - \mu_i})dx$$

$$= 2\int yI_1(2y)K_1(2y)dy = y^2\left[\left(1 + \frac{1}{4y^2}\right) I_1(2y)K_1(2y) \right.$$

$$\left. - I_1'(2y)K_1'(2y)\right] - \frac{1}{4}, \quad y \equiv \phi_{i-1}(x),$$

$$I_1'(2y) = 2\left[I_0(2y) - \frac{1}{y}I_1(2y) \right], \quad K_1'(2y) = -2\left[K_0(2y) + \frac{1}{y}K_1(2y) \right].$$

We emphasize that in Eq.(3.26) we don't need free constants since they already appear in zeroth-order solution. Therefore, the boundary conditions can be satisfied by choosing appropriate values for these constants.

Accordingly, the function $\Xi^{(1)}(x, \tau_j)$ can be found in a similar way. The overall solution is given by the expression Eq.(3.26), where $\phi_{i-1}(x)$ must be replaced with $\phi_i(x)$. This finalizes the construction of the first-order approximation.

We will not construct higher order approximations for $\hat{B}^{(s)}(X, \tau_j)$, $s > 1$ because incorporation of the first two terms already provides a good approximation with the accuracy of $O(\varepsilon^2)$ (since usually ε is of order 0.1 or less).

Also, as we observed in our numerical experiments, using these asymptotic solutions as part of the calibration procedure makes the latter fairly stable.

3.6 The calibration procedure

The calibration procedure runs sequentially for each time step beginning from $j = 1$ and up to $j = M$. Given the solution at the previous time step $B_{j-1}(X, \tau_j)$, we proceed by making some initial guess for the parameters $v_{j,i}^0, v_{j,i}^1,\ i = 0, \ldots, n_j$.[14] Actually, we need this guess just for $v_{j,i}^1$, $i = 0, \ldots, n_j$ and v_{j,n_j}^0, because based on Eq.(3.5)

$$v_{j,i}^0 = v_{j,n_m}^0 + \sum_{k=i+1}^{n_j} X_k(v_{j,k}^1 - v_{j,k-1}^1), \quad i = 0, n_j - 1. \tag{3.27}$$

So the total number of the unknown parameters to be determined is $n_j + 2$. Since for maturity T_j only n_j market quotes are given, we need two additional conditions to provide a unique solution. For instance, often traders have an intuition about the asymptotic behavior of the volatility surface at infinity, which, according to our construction, is determined by v_{j,n_j}^1 and $v_{j,0}^1$.

Using the analytical solution $\hat{B}_j(X, p)$ for a given maturity, the scaled Put option prices $B(X_i, \tau_j)$ can be calculated similarly to LS2011 by computing the inverse Laplace-Carson transform. The latter can be efficiently performed by using the Gaver-Stehfest algorithm

$$B(X, \tau_j) = \sum_{s=1}^{(N)} \frac{St_s^{(N)}}{k} \hat{B}(X, s\Lambda), \quad \Lambda = \frac{\log 2}{\tau_j}. \tag{3.28}$$

This algorithm was studied in many papers (see, e.g., [Kuznetsov (2013)] and references therein), and, provided that the resulting function is non-oscillatory, converges very quickly. For instance, choosing $N = 12$ is usually sufficient. The coefficients $St_s^{(12)}$ can be found explicitly, see, e.g., LS2011. It is also known that this algorithm requires high-precision arithmetic for its implementation. This effect is especially pronounced for small τ_j, so the inversion can become numerically unstable unless a sufficient number of significant digits is used.

Once all the option prices are computed, they can be compared with given market quotes. Hence, some kind of a least-square minimization procedure can be utilized to find the final values of all the unknown parameters

---

[14]If $j = 1$ the previous time solution is just the payoff function.

that fit model option prices to market quotes. Complexity-wise, at every iteration we need to compute the solution at n_j spatial points and N temporal points, the former are given, the latter are prescribed by the Gaver-Stehfest algorithm. Also every such solution, as it is defined in Eq.(3.12), requires $2n_j$ constants $C_1, ..., C_{2n_j}$, which solve the corresponding system of linear equations. As it was mentioned earlier, due to its special structure, this system can be solved with complexity of $O(2n_j)$. Overall, complexity of performing one iteration is $O(2n_jN)O(\text{Ku})$, where $O(\text{Ku})$ is complexity of computing all Kummer's functions for the solution in one spatial point. This seems to be a significant improvement in performance as compared, e.g., with LS2011, where computation of the source terms required numerical integration.

3.6.1 *Initial guess for the calibration*

Obviously, the calibration is a time-consuming process, therefore, having a smart initial guess significantly improves its convergence rate.

Suppose we have already obtained all values of the parameters for maturities T_j, $j \in [1, j_1]$, $j_1 < M$, and now need to run the calibration for the maturity T_m, $m = j_1 + 1$. Also suppose we are given market values $w(m, i)$, $i = 1, \ldots, n_m$ for the implied variance. To produce an "educated" initial guess for the calibration procedure, we suggest to use Eq.(3.2) to get the initial values of $v_{m,i}^1$, $i = 1, \ldots, n_m - 1$ and v_{m,n_m}^0. In particular, the first derivatives $\partial_T w(m, i), \partial_X w(m, i)$ in the right-hand-side of Eq.(3.2) can be approximated by the finite-differences of the first order using two given values of w in the strike and time space, and the second derivative $\partial_X^2 w$ — by using the second order approximation with three given values of w in the strike space. When computing $\partial_T w(m, i) \approx [w(m, K_i) - w(m - 1, K_i)]/\tau_m$ it is possible that market quote $w(m - 1, K_i)$ is not available at T_{m-1}; in this case interpolation/extrapolation in K over given quotes at T_{m-1} can be used to get this value. This calculation generate values for $\sigma_{m,i}$, $i \in \mathbb{Z} \cap [1, n_m]$. If some of them are negative, they can be replaced by a small positive number δ.

Next, we use Eq.(3.27) and obtain a system of linear equations of the form

$$v_{m,n_m}^0 + \sum_{k=i}^{n_m} X_k(v_{m,k}^1 - v_{m,k-1}^1) + v_{m,i}^1 X_i = \sigma_{i,m}, \qquad i \in [1, n_m - 1],$$

$$v_{n_m,m}^0 = \sigma_{n_m,m} - v_{n_m,m}^1 X_{n_m},$$

where the values v^1_{m,n_m}, $v^1_{m,0}$ are given. Since this system is upper triangular, it could be efficiently solved with linear complexity $O(n_m)$.

3.7   Option prices for short T

As was mentioned in the previous section, computation of the inverse Laplace transform by using the Gaver-Stehfest algorithm requires very high-precision arithmetic for small τ_j. Therefore, in this limit it does make sense to solve the modified Dupire's equation in a different way, namely by using an asymptotic expansion for its solution at $\tau_j \to 0$, see also LS2011.

For the time-homogeneous models of the local volatility, i.e., when the volatility does not depend explicitly on time, this problem was considered in various papers, see, e.g., [Gatheral *et al.* (2012)] and references therein. For the time-inhomogeneous model it was further analyzed in [Gatheral *et al.* (2012)]. In that paper an asymptotic representation for the European call option price $C(\tau_j, x)$ with $x = \log S$ was obtained by using an expansion of the transition density function of a one-dimensional time inhomogeneous diffusion. If $x < \log K$ this asymptotic solution reads

$$C^-(\tau_j, x) = \frac{v_j(K)K}{\sqrt{2\pi}} \frac{u_0(x, \log K)}{d^2(x, \log K)} \tau_j^{3/2} \exp\left[-\frac{d^2(x, \log K)}{2\tau_j}\right],$$

$$d(x, y) = \int_x^y \frac{dx}{\sqrt{v_j(x)}},$$

$$u_0(x, y) = \left(\frac{v_j(x)}{v_j^3(y)}\right)^{1/4} \exp\left[-\frac{1}{2}(y - x) + (r - q)\int_x^y \frac{ds}{v_j(s)}\right],$$

(3.29)

where the superscript $^{(-)}$ is used to indicate that this solution corresponds to $x < \log K$.[15] Also, when deriving Eq.(3.29) it is assumed that $\forall x \in \mathbb{R} \; \exists C > 0 : C^{-1} \le \sigma_j(x), |\sigma_j'(x)| \le C, |\sigma_j''(x)| \le C$. This assumption may fail at the boundaries when $S \to 0$ and $S \to \infty$.

Having Call prices $C(\tau_j, x_i)$ computed for all strikes K_i, $i = 1, \ldots, n_j$ and a particular maturity $\tau_j \ll \min_i(1/\sigma_{j,i})$, we can also compute the corresponding Put prices by using Call-Put parity. Then, running parameters for the local variance function can be found by calibration. Note, that since $v_j(x)$ is piecewise linear in X, the integral in Eq.(3.29) can also be constructed as a sum of various contributions. When calculating these contributions, we rely on the fact that if x, y belong to the interval i, the

---

[15]In our notation $x = \log K - (r - q)T - X$.

variance on this interval is given by Eq.(3.4), so that

$$d(x, y) = \int_x^y \frac{dx}{\sqrt{b - a_2 x}} = \frac{2}{a_2}\left(\sqrt{b_2 + a_2 Y} - \sqrt{b_2 + a_2 X}\right),$$

$$\int_x^y \frac{ds}{v_j(s)} = \frac{1}{a_2}\log\frac{b_2 + a_2 Y}{b_2 + a_2 X},$$

where $b = b_2 + a_2(\log K + \kappa)$.

If $x > \log K$, from [Gatheral *et al.* (2012)] we have

$$C^+(\tau_j, x) = e^x - Ke^{-r\tau_j} - C^-(\tau_j, x).$$

3.8   Results and discussion

In our numerical test we use the same data set as in [Itkin (2015)], i.e., we take data from http://www.optionseducation.org on XLF traded at NY-SEArca on March 25, 2014. The spot price of the index is $S = 22.64$, and $r = 0.0148$, $q = 0.01$. The option implied volatilities (IV) are given in Tables 3.2,3.3. We take all OTM quotes and some ITM quotes which are very close to the at-the-money (ATM).

When strikes for Calls and Puts coincide, we take an average of I_{call} and I_{put} with weights proportional to $1 - |\Delta|_c$ and $1 - |\Delta|_p$ respectively, where Δ_c, Δ_p are option Call and Put deltas.[16]

We have already mentioned that in our model for each term the slopes of the smile at plus and minus infinity, v^1_{j,n_j} and $v^1_{j,0}$, are free parameters. So often traders have an intuition about these values. However, in our numerical experiments we take for them just some plausible values, which are given in Table 3.5.

When calibrating the model to market data, we use the standard Matlab *fmincon* function. We start by using an "active-set" algorithm (see Matlab documentation on *fmincon*), and if it doesn't converge, switch to an "sqp" algorithm (it is also described in the Matlab documentation). We emphasize that optimization of this step is not a subject of this research, and for a more detailed discussion of various problems related to the calibration of the local volatility surface we refer the reader to a recent paper [Lindholm (2014)] and references therein.

---

[16]By doing so we do take into account effects reported in [Ahoniemi (2009)], who pointed out that the IVs calculated from Call and Put option prices corresponding to the same strike do not coincide, although they should be equal in theory. Our weights are chosen according to a pure empirical rule of thumb, and a more detailed investigation of this effect is required.

Table 3.2: XLF implied volatilities for the Call options.

T	K, Put						
	18	19	20	21	21.5	22	23
4/4/2014	-	-	39.53	23.77	19.73	16.67	-
4/19/2014	-	32.90	26.79	20.14	-	15.19	12.93
5/17/2014	33.27	26.88	23.08	18.94	-	16.12	13.86
6/21/2014	27.84	23.90	21.07	18.88	-	16.95	15.82
7/19/2014	26.09	22.81	20.29	18.13	-	16.30	14.93
9/20/2014	24.20	22.23	20.32	18.76	-	17.40	16.41

Table 3.3: XLF implied volatilities for the Put options.

T	K, Call									
	21	21.5	22	22.5	23	24	25	26	27	28
4/4/2014	-	16.60	14.69	14.40	14.86	-	-	-	-	-
4/19/2014	-	-	15.79	-	13.38	15.39	-	-	-	-
5/17/2014	16.71	-	14.48	-	-	13.75	-	-	-	-
6/21/2014	16.31	-	14.78	-	-	13.92	14.28	16.58	-	-
7/19/2014	16.82	-	15.24	-	-	14.36	14.19	15.20	-	-
9/20/2014	17.02	-	15.84	-	-	14.99	14.56	14.47	14.97	16.31

Method	$T \ll 1$	$|a| \gg 1$	general
Time, sec	1.0-1.4	1-7	5-7

Table 3.4: Typical time to converge (per strike) using various algorithms for computing $\hat{B}(X, T)$.

Table 3.5: Parameters $v^1_{j,0}$ and v^1_{j,n_j} for the option data in Tables 3.2 and 3.3.

j	T_j	$v^1_{j,0}$	v^1_{j,n_j}
1	4/04/2014	-0.1206	0.1000
2	4/19/2014	-0.1000	0.1000
3	5/17/2014	-0.1309	0.1000
4	6/21/2014	-0.1000	0.1000
5	7/19/2014	-0.1000	0.1000
6	9/20/2014	-0.1000	0.1000

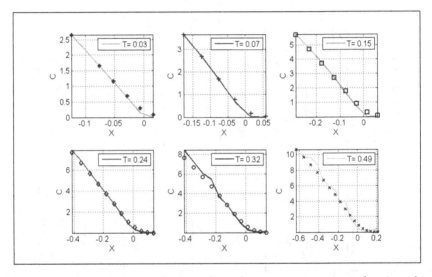

Figure 3.3: Term-by-term fitting of market prices constructed using the whole set of data in Tables 3.2 and 3.3.

To elaborate a bit more on this point, construction of the local volatility surface given the market data on vanilla European options requires solving two embedded problems. One is to provide an efficient optimization algorithm to solve, e.g., a minimization problem which appears if one uses a least-square approach. There is a wide literature on various approaches to solving this problem, which is known to be ill-posed. The other problem is that when running this optimization, at every step we need to compute either the theoretical option prices or the corresponding implied volatilities, by solving the Dupire equation. Here we deal with the second problem by (i) given T we use a piecewise linear approximation in X of the local variance term, and (ii) provide a no-arbitrage interpolation of the source term in X[17] which allows the whole solution to be obtained in closed form.

Therefore, since we don't consider the optimization in detail, it is provided here for illustrative purposes, and, certainly, a more sophisticated and powerful algorithm could be used to a greater effect.

The results of such a calibration which is done term-by term, are given in Fig. 3.3. Here each subplot corresponds to a single maturity T (marked in the legend) and shows market data (discrete points) and computed values (solid line). This simple local calibration algorithm provides rather decent results, except for the vicinity of $X = -0.5$ in the last subplot.

For the first two maturities we successfully use the asymptotic method described in Section 3.7. Then, for the next two maturities, the method described in Section 3.5 provides good results. Finally, for the last two maturities a combination of the general algorithm with that described in Section 3.5 has to be used.

The local variance curves obtained as a result of this fitting are given term-by-term in Fig. 3.4. The corresponding local variance surface is represented in Fig. 3.5

It can be seen that the local variance is positive everywhere on the grid, so that our construction is arbitrage-free.

Performance-wise the proposed algorithm is reasonably efficient. Indeed, we ran our tests in Matlab using two Intel Quad-Core i7-4790 CPUs, each of 3.80 Ghz. As was mentioned in the previous section, the calibration time strongly depends on the method chosen to compute $\hat{B}(X, T)$. Typical results are given in Table 3.4. These results are normalized per number of strikes for a given term. Obviously, they could be considered just as a crude estimation, since the convergence strongly depends on the quality of the initial guess. In our calculations we used the approach described in Section 3.6.1. Still, it can be seen that the second method in Table 3.4 is slower than the first one as it requires the evaluation of the Bessel functions. The third method requires multiple computations of Kummer's functions and is the slowest one. However, as we use the Gaver-Stehfest algorithm, it can be fully parallelized. Same is true for the computation of Kummer's functions in all points X_i, $i \in \mathbb{Z} \cap [1, n_j]$ for a given maturity T_j, which we do at every iteration of the calibration procedure. Therefore, having a sufficient

---

[17]The no-arbitrage in T, i.e., the calendar no-arbitrage is already addressed in this approach since we use an exact in time solution of the Dupire equation given by the inverse Laplace transform.

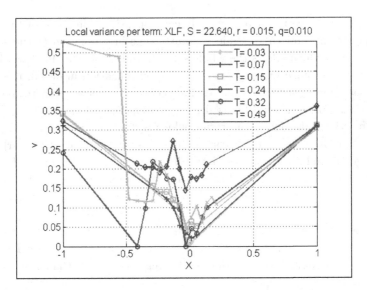

Figure 3.4: Term-by-term fitting of the local variance.

number of cores, a potential speedup of the parallel implementation should be proportional to $N = 12$ (the number of the Gaver-Stehfest algorithm time steps) times the number of strikes. In our case this provides the calibration in less than a second per maturity even when the general method is used.

To summarize, in this chapter we described the methodology of constructing a semi-analytical solution of the Dupire equation proposed in [Itkin and Lipton (2018)], and construction of the local volatility surface based on this methodology. This approach extends the approach proposed in LS2011 by replacing a tiled local variance shape with a piecewise linear construction and relaxing their assumptions about zero interest rates and dividend yields. Yet, our approach, which combines an application of the Laplace-Carson transform and solution of the resulting inhomogeneous ordinary differential equation in terms of Kummer's hypergeometric functions, remains analytically tractable.

When solving the modified Dupire equation by utilizing the Laplace-Carson transform method, one must be cognizant of the following issue. To compute the source term $h(X) = pI_{12}(X)$ at some time step j we need the function $B_{j-1}(X, \tau_{j-1})$ obtained at the previous time step. However, the market quotes for the maturities T_j and T_{j-1} could be given at different sets

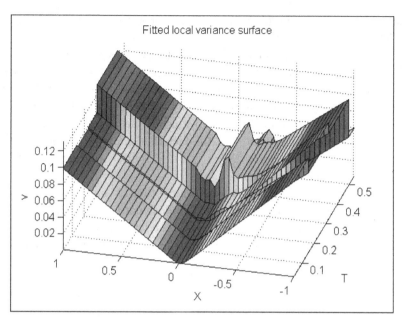

Figure 3.5: The local variance surface constructed using the proposed approach.

of X even if the strikes K are same, since, by definition, $X = \log(K/F)$ and $F = F(T)$. Therefore, we need the values of $B_{j-1}(X, \tau_{j-1})$ at certain points X where they have not been calculated yet. LS2011 used an interpolation to obtain the required values. However, this interpolation must be carefully constructed to preserve no-arbitrage, and this problem was not addressed in LS2011. Here we propose an interpolation which allows computation of the source terms in closed form (while in LS2011 an additional numerical integration for computing the source terms was required), and prove that our interpolation does not create arbitrage. Overall, our approach is more accurate (piecewise linear term instead of a piecewise constant), more efficient (closed form solution instead of a numerical integration) and more reliable (proven no-arbitrage).

In addition, we noticed that using the general algorithm for small maturities or steep local variance slopes often results in various inefficiencies and instabilities. Therefore, for these special cases we propose alternative methods constructed by using asymptotic (regular or singular) expansions, which do not suffer from these issues. In our opinion, this is an interesting and practically important extension of the general methodology described in the previous two paragraphs.

The numerical experiments demonstrate robustness of our approach. Obviously, closed-form solutions for the source terms and asymptotic solutions, expressed in terms of functions less computationally expensive than Kummer's functions, significantly speed up the calibration. The implementation could be made more efficient by using the internal parallelism of the Gaver-Stehfest algorithm, and the fact that Kummer's functions corresponding to different points X_i, $i \in \mathbb{Z} \cap [1, n_j]$ for a given maturity T_j could be computed in parallel.

By its nature, our model (as well as any other LV model) provides just a fit for the current market snapshot, and does not consider any dynamics for the local volatility surface *itself*. While the latter issue should be investigated separately, our choice of the LV is parsimonious enough to greatly facilitate this endeavor.

Chapter 4

Regression-based Methods

In this chapter we describe the second and, perhaps, the most popular approach to building the local volatility surface by regressions. Regression-based methods include both parametric and non-parametric fits. Usually, all these methods deal with construction of the implied volatility surface while the local volatility can be found afterwards by using Eq.(3.2) or any its flavor.

4.1 Overview

The amount of literature on constructing the implied volatility surface is huge, and this topic can be subject of a separate book. Obviously, we cannot describe all available ideas and approaches within one chapter. Therefore, we consider only the most popular models while trying to refer the reader to the existing surveys where they are available. From this prospective, for the survey of methods published before 2014 we mention [Gatheral (2006); Homescu (2011, 2014)] and references therein. More recent books include [Bergomi (2016); Derman *et al.* (2016)].

Having said that, below we provide just a short description of the main ideas related to construction of the parametric regressions. As the examples of this methodology, in the next Sections we describe a very popular SVI model of J. Gatheral, and also the parametric model proposed in [Itkin (2015)].

In general, various parameterizations of the implied volatility (IV) surface were proposed to address several goals:

(1) Given an European option and a set of market quotes for various (K, T), construct an arbitrage-free local volatility surface. Then, for instance, it could used as an input for calibration of more sophisticated local stochastic volatility model to cover both exotic and European options.

(2) Use the IV surface for pricing OTC options and other derivatives with strikes and maturities other than that offered by the option exchanges.

(3) Assess an adequacy of an option pricing model based on the shape of the IV surface.

(4) Option traders and marker makers often use the current snapshot of the implied volatility over all strikes and maturities as a basis to produce a short-term volatility forecast over future periods of time using some assumptions about the future dynamic of the IV.[1]

There exist two major approaches to construct an arbitrage-free IV surface. The first one uses some stochastic model for the underlying spot or forward price which is calibrated to the market data. For instance, in the Equity or Fixed Income worlds one can use the popular Heston ([Heston (1993)]) or SABR ([Hagan *et al.* (2002)]) model, calibrate it to the market data and then use this model to find the IVs for the missing strikes and expirations where the market quotes are not available. By construction the IVs produced by the arbitrage-free model are also arbitrage-free. However, the main problem with this approach is that it is difficult to come up with a model which is rich enough to fit well the observed market data.

Another approach does not consider any model of the underlying, but instead uses some parametric fit of the implied volatility surface. Parametric models of the IV came to a regular consideration at the end of 1990s. Several parametric models for the IV surface were suggested by [Dumas *et al.* (1998)],[2] and adapted and tested for FTSE options by [Alentorn (2004)] In the Dumas parametric model the IV surface is modeled as a quadratic function of the so-called normalized strike (rather than the strike price). Later this approach was further extended by [Tompkins (2001);

---

[1]As was mentioned by one of referees, a single point on the implied volatility surface could potentially be such a forecast. Also market models of implied volatility, e.g., [Cont and Fonseca (2002)] tell us that implied volatilities also forecast their covariance with spot and their own volatilities as well. However, the usefulness of even a single implied volatility as a forecast is hampered by the difference between risk-neutral and real world probability measures. It is well known that for S&P500, the at-the-money forward implied volatility is on average above the subsequent realized volatility, suggesting that this distinction is important and empirically verifiable. However, in a short run, say up 10 minutes, in a quiet market such a forecast could be potentially helpful.

[2]He actually suggested a model for the local volatility, which, however, could be re-mapped to the implied volatility.

Kotzé *et al.* (2013); Carr *et al.* (2013)]. The normalized strike is defined as

$$z = \frac{\log(K/F)}{\sigma_* \sqrt{T}}, \tag{4.1}$$

where K is the option strike, F is the forward price, T is the time to expiration, and σ_* is the normalization constant which usually is set either to 1, or to the ATM implied volatility. The normalized strike is a unit-less quantity. Some people also call it moneyness or log-moneyness, however we reserve this word for a standard definition of the forward moneyness as $M = K/F$. By definition normalized strike vanishes at the forward money (ATM). For a call option, positive normalized strike corresponds to the In-The-Money option and negative normalized strike — to the Out-of-The-Money option. Usually the normalized strike is used under an assumption of "sticky moneyness" which means that the IV doesn't change when z stays constant (it is also known as "sticky delta"), which allows elimination of refitting the volatility smile within some postulated period of time even when the underlying price changes. This is different from another popular assumption which is called a "sticky strike" rule ([Derman and Kani (1994b); Derman (1999); Sinclair (2013)]).

Despite within the second approach (a static parameterization) the quality of the fit is often better than in the first one (a dynamic model of the underlying or the implied volatility itself), static parameterization tells us something just about the current market snapshot of the option prices/IVs, and nothing about the temporal dynamics of the IV. For instance, the above mentioned "sticky" assumptions about the future dynamics of the IVs are irrelevant to the parameterization itself. Clearly one can construct such a parameterization using the normalized strike as a convenient underlying variable if he/she relies on a "sticky log-moneyness" dynamics assumption to be true. This, however, doesn't mean that another parameterization which uses the normalized strike as an underlying variable and relies on a sticky-moneyness assumption might not be used to fit the same set of the market IVs. That is because this type of parameterizations is static by nature. In other words, it is impossible to forecast the future IVs *per se* using this static fit. Rather, when using this approach by term "forecasting" the practitioners usually mean that when the IV of some option with time horizon (maturity) T is known, it provides some average value of volatility from today to T. This, however, is not a property of the parameterization, but rather the property of the current option market to provide some "on average" information about the future behavior of the stock market.

An extended work on modeling the IV surface using the static approach has been done by Gatheral in many papers, starting perhaps with [Gatheral (2004)]). He used a different parameterization of the smile, known as stochastic-volatility-inspired (SVI) model, which is driven by a forward log-moneyness $\chi = \log(K/F)$. Gatheral and co-workers also proposed some empirical dependencies of how parameters of the fit evolve with time, [Gatheral (2006)]. Later in [Gatheral and Jacquier (2011)] it was shown that the SVI parameterization and the large-time asymptotic of the Heston implied volatility agree algebraically, which provides an additional theoretical justification for the above parameterization. Finally, a no-arbitrage version of the SVI model was proposed in [Gatheral and Jacquier (2014)]. We describe this approach in more detail in Section 4.2.

Some other static parameterizations were also proposed in the literature, e.g. [Fengler (2005); Zhao and Hodges (2013); Andreou *et al.* (2014); Sehgal and Vijayakumar (2008); Daglish *et al.* (2007); Carr *et al.* (2013); Romo (2011); Rosenberg (2000)], and also references therein. In Section 4.3 we discuss the main highly desirable features that any such a parameterization should provide the user with. It could be observed that in contrast to the old approaches, recent models, e.g. the extended SVI model, and models in [Kotzé *et al.* (2013); Zhao and Hodges (2013)] do make account for these features, and thus could be useful in practice.

As far as a demand for the dynamic models of the IV is concerned, [Cont and Fonseca (2002)] considered the prices of the index options at a given date (they are usually represented via the corresponding IV surface) that clearly demonstrated skew/smile features and also a term structure, the behavior that several IV models have attempted to reproduce. They underlined that the IV surface also changes dynamically over time in a way that is not taken into account by the existing modeling approaches, giving rise to a Vega risk in option portfolios. Using time series of option prices on the S&P500 and FTSE indices, they studied the deformation of this surface and showed that it may be represented as a randomly fluctuating surface driven by a small number of orthogonal random factors. Then Cont and Fonseca identified and interpreted the shape of each of these factors, studied their dynamics and their correlation with the underlying index. A simple factor model compatible with the empirical observations was proposed. The authors illustrated how this approach simulates and improves the well-known "sticky moneyness" rule used by option traders for updating the IVs. Their approach gave justification for using Vega when measuring the volatility risk, and provided decomposition of the volatility risk as a sum of contributions from empirically identifiable factors.

It is worth mentioning that the popular assumptions of "sticky strike" or "sticky moneyness" are just an empirical rule-of-thumb. For instance, [Ciliberti *et al.* (2008)] analyzed these assumptions by considering in detail the skew of some stock option smiles, which is induced by the so-called leverage effect on the underlying, i.e., the correlation between past returns and future square returns. This naturally explains the anomalous dependence of the skew as a function of the option maturity. The market cap dependence of the leverage effect is analyzed using a one-factor model. The authors show how this leverage correlation gives rise to a non-trivial smile dynamics, which turns out to be intermediate between the "sticky strike" and the "sticky delta" rules. Finally, they compare their result with stock options data, and find that the option markets overestimate the leverage effect by a large factor, in particular, for the long-dated options. This subject requires some further investigation.

Another interesting idea was proposed in [Carr and Wu (2010)]. This paper considers the future dynamics of the Black-Scholes implied volatility surface, and derives no-arbitrage constraints on the current shape of the volatility surface. Under the specified proportional volatility dynamics, the shape of the surface can be cast as solutions to a simple quadratic equation. Furthermore, corresponding to the option implied volatility for each contract, the paper defines a new, option-specific expected volatility measure that can be estimated from the historical sample price path of the underlying security. The measure is defined as the volatility input that generates zero expected delta-hedged gains from holding this option and can thus differ across different option strikes and expiries. Applying the new theoretical framework to the S&P500 index options market, the authors extract volatility risk and volatility risk premium from the two volatility surfaces, and find that the extracted volatility risk premium significantly predicts future stock returns. Thus, knowledge of the future dynamics also eliminates the necessity in any artificial assumptions like "stickiness", etc. See, a recent paper of [Sepp (2014)] and also [le Roux (2007); Romo (2014)].

So far, most of the IV researchers have been focused on Equity and FX derivatives. However, [Borovkova and Parmana (2009)] applied this idea to the option price data from oil markets. They combined the simplicity of the Gatheral parametric method with the flexibility of a non-parametric approach. The authors claim that the method can successfully deal with a limited amount of the option price data. Performance of the method was investigated by applying it to prices of the exchange-traded crude oil and gasoline options, and the results were compared with those obtained by a purely parametric approach. Furthermore, investigation of the relationship between volatilities implied from the European and Asian options showed that the Asian options in oil markets are significantly more expensive than the theoretical arguments imply.

To summarize, various static parameterizations were in use by traders since 1990 when the skew became pronounced in the market. However, as practitioners observed in their day-to-day trading, even the best models such as SVI and recent versions of the quadratic fit sometimes fail to fit well the market data. The author's own experience also justifies a failure to fit these models to the data sets, obtained from some data providers. Also, according to [Biscamp (2008)] the SVI model was thoroughly tested by practitioners in recent years and did not prove to work well for all products (like the index options, dispersion, equity options etc.). Therefore, some trading firms run their own proprietary models that exploit an idea of building a piecewise polynomial smile in the z space. This approach also has some problems, namely:

- Determining a boundary point between two pieces of the smile, where in addition the smile is C_2 continuous. Usually it requires solving some non-linear equation, which is expensive. The necessity of solving the nonlinear equation slows down the volatility smile fit, and especially computation of derivatives of the smile with respect to the model parameters which usually are computed by the bump-and-grind method.
- This approach still does not resolve the problem of fitting maturities close to expiration.
- This functional form does not fit the market data well for both skew and smile.
- The asymptotic behavior of the smile at wings in z does not agree with the result of [Lee (2004)] that the variance should be asymptotically linear in z.

All the above suggests that a model suitable to better fit the static market volatility data could be helpful. Motivated by this, in [Itkin (2015)] a rather flexible model was proposed that we further discuss in Section 4.3. The model amounts to resolving the discussed issues with the existing approaches. We also show how to construct a arbitrage-free IV surface by using an arbitrage-free interpolation and/or extrapolation if necessary.

We emphasize that according to [Carr (2014b)] any such a formula must provide the following three properties:

(1) It analytically describes implied volatilities instead of option prices.
(2) It exactly fits any set of arbitrage-free mid-market implied volatilities.
(3) It does not produce arbitrage.

Similar thoughts could be found in [Rebonato (2004); Castagna (2010)].

While the first one is obvious, it is usually hard to guarantee the last two properties. In [Itkin (2015)] an exact fit to the given mid-market quotes is also not guaranteed since the fit is provided by using a least-square optimization. However, that approach does guarantee, that the regressed implied volatility is in between of the given bid and ask, and is close (in some norm) to the mid price. Second, by construction it guarantees no-arbitrage in time and for a given grid of strikes. This grid could be non-uniform, and it consists of the nodes with different strikes and/or maturities provided by the user. The intermediate quotes could be found by the arbitrage-free interpolation/extrapolation in the strike space as this is described in Chapter 2.

As an example, suppose we use the local stochastic volatility model to price and hedge a set of exotic and vanilla options simultaneously. To do that we need a local volatility (LV) surface calibrated to the market data. The appropriate LV grid could, e.g., coincide with the finite-difference grid in the spot space. To get the LV surface we may first build the IV surface and calibrate it to the vanilla quotes, and the use the Dupire's formula to re-map the IV into the LV. When using such an approach we are not interesting in the values of the implied volatilities in between the grid nodes, and, therefore, the proposed method could be applied. We also guarantee a correct asymptotic behavior of the smile at both large positive and negative normalized strikes.

The latter means that this model of the implied volatility is a *discrete* model defined at a given given set of "states" (strikes), similar to, say, a discrete Markov chain model. And we are not aware of any continuous limit of this model at the moment. Compare this with the SVI model where a nice result is available that the model structurally coincides with the hight T asymptotic of the Heston model.

At the end, we have to mention that recent progress in the field of artificial intelligence (AI), machine learning (ML) and also in computer industry resulted in the ongoing boom of using these techniques as applied to solving complex tasks in science and industry also including the financial industry and mathematical finance. This technique could be used for modeling the implied volatility surface as well. For instance, in a recent paper

[Zheng *et al.* (2019)], unlike some previous studies, in which machine learning algorithms were used directly as a "black box", the authors propose an approach which tailored to the implied volatility surfaces. This means that they first construct a so-called gated deep neural network, and then calibrate (train) it by incorporating the related financial conditions and empirical evidence such as no static arbitrage,[3] boundaries, asymptotic behavior of slopes, etc. The latter is done in the form of soft constraints or a penalty function. The training set consists of the option data on the S&P 500 index over twenty years (i.e., the historical prices), and even includes the options with a short time to maturity. The paper claims that this approach outperforms the widely used SVI model on the mean average percentage error in both in-sample and out of-sample datasets. It also outperforms other similar neural network models which do not incorporate the financial conditions and empirical evidence.

The main advantage of this approach is that it doesn't require any particular parametric form of the smile (despite a particular activation function is proposed in [Zheng *et al.* (2019)]). Rather, it relies on a full power of deep learning technique which uses non-linear regressions to fit the data subject to the necessary constraints. Therefore, this approach has a lot of advantages. At the same time, as the financial constraints are introduced in the form of penalty functions, this doesn't guarantee the exact fulfillment of these conditions, even for in-sample data. Obviously, for out of-sample data one can only believe that these conditions are still satisfied. In more detail this problem is discussed in the recent paper of the author, [Itkin (2019)]. Despite being very interesting, we don't have space to discuss the deep learning approach in detail in this book.

4.2 SVI model

The *stochastic volatility inspired* (or SVI) parameterization was developed by J. Gatheral in 1999 at Merrill Lynch and then publicly presented in [Gatheral (2004)]. The model was inspired by earlier parameterization of the implied variance (rather than the implied total variance) invented by Tim Klassen at Goldman Sachs. The construction was motivated by two key features the model should follow:

- As follows from Roger Lee's formula [Lee (2004)], the implied Black-Scholes variance should be linear in the log-strike χ at fixed T. In other

---

[3]However, they missed no-arbitrage conditions for the so-called vertical spread.

words, the implied variance as a function of χ at wings $\chi \to \pm\infty$ is linear with the slope $0 < \phi(\infty) < 2$.[4]

- The parameterization should be simple enough to provide a fast calibration to market quotes. At the same time it should reproduce both smile and skew of the listed options.

According to [Gatheral (2004)], the SVI parametrization is of the form

$$w(\chi; a, b, \sigma, \rho, m) = a + b\left\{\rho(\chi - m) + \sqrt{(\chi - m)^2 + \sigma^2}\right\}. \quad (4.2)$$

Here w is the total implied variance, $w = \sigma_{BS}^2 T$, σ_{BS} is the implied volatility, $\chi = \log(K/F)$, $a \in \mathbb{R}$ gives the overall level of variance, $b \geq 0$ gives the angle between the left and right asymptotes, $\sigma > 0$ determines how smooth the vertex is, $|\rho| < 1$ determines the orientation of the graph, and changing $m \in \mathbb{R}$ translates the graph. Obviously, these parameters should obey the condition $a + b\sigma\sqrt{1 - \rho^2} \geq 0$ which ensures that $w(\chi; a, b, \sigma, \rho, m) \geq 0$ for all $\chi \in \mathbb{R}$.

As per [Gatheral and Jacquier (2014)] changes in the parameters of the model have the following effects:

- Increasing a increases the general level of variance, a vertical translation of the smile;
- Increasing b increases the slopes of both the put and call wings, tightening the smile;
- Increasing ρ decreases (increases) the slope of the left(right) wing, a counter-clockwise rotation of the smile;
- Increasing m translates the smile to the right;
- Increasing σ reduces the at-the-money (ATM) curvature of the smile.

Despite the parameterization in Eq.(4.2) is simple, it is not intuitive to traders. This is inherent not just to SVI, but almost to all known parameterizations. Also, suppose that we calibrate this model to a current snapshot of the option quotes. Since the fit is static, the model should be re-calibrated later at some time when the trader believes the market moves are significant enough, and the previous fit cannot be used. Therefore, a natural question would: how stable are parameters of the fit over various calibrations. This questions is also considered later in Section 4.2 as

---

[4]Note, that an asymptotic no-arbitrage argument was pioneered by [Hodges (1996)], then [Gatheral (1999)] and later [Lipton (2001)] who mentions that the resulting $IV(\chi)$ bounds are $O(|\chi|^{1/2})$ for large $|\chi|$. This was then further extended by the familiar results of [Lee (2004)].

applied to the model of [Itkin (2015)]. In general, there is no reason to expect these parameters of the fit to be particularly stable.

Another, more intuitive version of the SVI model is called the SVI-Jump-Wings (SVI-JW), and is parameterization of σ_{BS}. For a given $T > 0$ and a parameter set $\xi_J = \{v_T, \psi_T, p_T, c_T, \tilde{v}_T\}$ the SVI-JW parameters are defined from the SVI model as follows, [Gatheral and Jacquier (2014)]

$$v_T = \frac{a + b\left(-\rho m + \sqrt{m^2 + \sigma^2}\right)}{T}, \tag{4.3}$$

$$\psi_T = \frac{b}{2\sqrt{w_T}}\left(\rho - \frac{m}{\sqrt{m^2 + \sigma^2}}\right),$$

$$p_T = \frac{b}{\sqrt{w_T}}(1 - \rho),$$

$$c_T = \frac{b}{\sqrt{w_T}}(1 + \rho),$$

$$\tilde{v}_T = \frac{1}{T}\left(a + b\sigma\sqrt{1 - \rho^2}\right),$$

with $w_T = v_t T$. This parameterization has an explicit dependence on T, and hence can be viewed as generalizing the raw SVI parameterization which is independent on expiration T. The SVI-JW parameters have the following interpretations:

- v_T gives the ATM variance;
- ψ_T gives the ATM skew;
- p_T gives the slope of the left (put) wing;
- c_T gives the slope of the right (call) wing;
- $\tilde{v}_T$ is the minimum implied variance.

In [Gatheral and Jacquier (2014)] it is argued that in practice the observed smiles are scaled almost perfectly as $1/\sqrt{w_T}$. If so, the parameters in Eq.(4.3) are constant and independent of the slice t. This is a convenient feature as it makes it easy to extrapolate the SVI surface to expirations beyond the longest expiration in the data set.

The inverse map from the SVI-JW parameters to those of the SVI is given by Lemma 3.2 in [Gatheral and Jacquier (2014)] which we present below for completeness:

Lemma 4.1 ([Gatheral and Jacquier (2014)]). *Suppose $m \neq 0, T > 0$ and define the quantities*

$$\beta = \rho - \frac{2\psi_T\sqrt{w_T}}{b}, \quad \alpha = \frac{\text{sign}(\beta)}{\beta}\sqrt{1 - \beta^2}.$$

Further assume that $\beta \in [-1, 1]$ *which is true if* $-p_T \leq 2\psi_T \leq c_T$. *Then the raw SVI and SVI-JW parameters are related as follows:*

$$b = \frac{\sqrt{w_T}}{b}(c_T + p_T),$$

$$\rho = 1 - p_T \frac{\sqrt{w_T}}{b},$$

$$a = \tilde{v}_T T - b\sigma\sqrt{1 - \rho^2},$$

$$m = \frac{(v_t - \tilde{v}_T)T}{b}\left(-\rho + \text{sign}(\alpha)\sqrt{1 + \alpha^2} - \alpha\sqrt{1 - \rho^2}\right)^{-1},$$

$$\sigma = \alpha m.$$

If $m = 0$, *this map still holds, but* $\sigma = (v_T T - a)/b$.

So far we discussed just various parameterizations inherent to the SVI model. However, they should be supported by some conditions on the parameters which would guarantee no-arbitrage. This conditions are also established in [Gatheral and Jacquier (2014)]. For instance, for the raw SVI parameterization a sufficient condition for the absence of calendar spread arbitrage reads:

Lemma 4.2 (Lemma 3.3 in [Gatheral and Jacquier (2014)]). *By definition, there is no calendar arbitrage if for any two expires* $T_1 \neq T_2$ *the corresponding slices* $w(K, T_1), w(K, T_2)$ *do not intersect. Let these two slices be characterized by the sets of the SVI parameters* $\xi_1 := \{a_1, b_1, \sigma_1, \rho_1, m_1\}$ *and* $\xi_2 := \{a_2, b_2, \sigma_2, \rho_2, m_2\}$. *The raw SVI surface in Eq.(4.2) is free of calendar spread arbitrage if a quartic polynomial* $\sum_{i=0}^{4} \alpha_i \chi^i$ *has no real root. Here, coefficients* α_i, $i \in [0.4]$ *can be explicitly expressed in terms of the parameters in* ξ_1, ξ_2, *and could be found on http://faculty.baruch.cuny.edu/jgatheral.*

The proof and further details could be found in [Gatheral and Jacquier (2014)].

To consider static arbitrage, a concept of SSVI surface is introduced, which is defined as

$$w(\xi, T) = \frac{\theta_T}{2}\left\{1 + \rho\phi(\theta_T)\chi + \sqrt{(1 - \rho^2) + [\rho + \phi(\theta_T)\chi]^2}\right\}, \qquad (4.4)$$

where $\theta_T = \sigma_{BS}^2(0, T)$ is the at-the-money (ATM) implied total variance, ϕ is a smooth function from $\mathbb{R}_+^*$ to $\mathbb{R}_+^*$ such that the limit $\lim_{T \to 0} \theta_T \phi(\theta_T)$ exists in $\mathbb{R}$. This representation amounts to considering the volatility surface in terms of ATM variance time, instead of standard calendar time. It should also be assumed that $\theta \in \mathcal{C}^1$ on $\mathbb{R}_+^*$, and $\theta_0 = 0$ an ATM option with zero time to expiry has no value.

The following Theorems are proved in [Gatheral and Jacquier (2014)].

Theorem 4.1. *The SSVI volatility surface in Eq.(4.4) is free of butterfly arbitrage if the following conditions are satisfied for all $\theta > 0$:*

$$\theta\phi(\theta)(1 + |\rho|) < 4,$$
$$\theta\phi^2(\theta)(1 + |\rho|) \leq 4.$$

Theorem 4.2. *The SSVI volatility surface in Eq.(4.4) is free of calendar spread arbitrage if and only if*

$$\partial_T \theta_T \geq 0, \ \forall T \geq 0,$$

$$0 \leq \partial_\theta(\theta\phi(\theta)) \leq \frac{1 + \sqrt{1 - \rho^2}}{\rho^2}\phi(\theta),$$

where the upper bound is infinite when $\rho = 0$.

Finally, as the SVI construction is done by using a discrete set of expirations, it is not immediately obvious how to interpolate these smiles (in case one needs it for the intermediate expirations) in such a way as to ensure the absence of static arbitrage. In [Gatheral and Jacquier (2014)] such an interpolation in time is proposed that guarantees no-arbitrage in time. We omit a detailed consideration of this problem here because (i) we have already discussed no-arbitrage interpolations in Chapter 2, and (ii) in Section 4.3.6 we will discuss it as applied to the model [Itkin (2015)] using a similar approach.

4.3 Model of [Itkin (2015)]

4.3.1 *Motivation*

Before we describe our construction, it is interesting to note that traditional parametric models represent the smile as some polynomial function of z. One of the reasons for doing this is that according to [Cont and Fonseca (2002)] the IV patterns across moneyness vary less in time than when expressed as a function of the strike. Also, there is an additional computational benefit by regressing at moneyness rather than at the strike prices, since the function is of a simpler form, and, therefore, the estimation algorithm converges faster.

A typical study is that of [Alentorn (2004)] where using data in the FTSE 100 index, the following models were tested:

$$\sigma(z) = \beta_0 + \beta_1 z + \beta_2 z^2 + \epsilon \qquad (4.5)$$
$$\sigma(z) = \beta_0 + \beta_1 z + \beta_2 z^2 + \beta_3 T + \beta_4 T z \epsilon,$$

where β_i, $i \in [0, 4]$ are the regression parameters that usually are a function of time, therefore Eq.(4.5) with the fixed coefficients represents just one term $T = const$ of the volatility surface. In [Borovkova and Parmana (2009)] the authors use a similar regression. They also noticed that the parabolic shape of the implied volatility function for a fixed maturity is the average shape of the actual volatility functions. Note that increasing the power of the polynomial volatility function (from two to three or higher) does not really offer a solution here, since this volatility function will still be the same for all maturities. Quadratic profile of the implied volatility as a function of z is also supported by PCA analysis of the implied volatility surface ([Cont and Fonseca (2002); Alexander (2001); Fengler *et al.* (2003)]).

Comparison of these models with the market data showed that they are able to capture a form of the volatility smile in the ATM region while often fail at wings. Another problem is fitting the smile close to expiration. Here $T \to 0$ implies $z \to \infty$, and the volatility at wings tends to infinity which is not supported by the market data. Therefore, the regression coefficients β_1, β_2 must tend to zero, and the fitting function degenerates in this limit. This poses a real problem for the optimization routine (it never converges to such a limit).

In Section 4.2 we described the SVI parametrization of J. Gatheral. To recall, in [Gatheral (2004)]) he derives necessary and sufficient conditions for the IV surface to be arbitrage free and shows how this parametrization fits the IV surfaces generated by various currently popular models, including the stochastic volatility and jump models. Also some examples are provided where the SVI well fits the actual IV surfaces — even the notoriously hard-to-fit very short expirations. Later, it was observed that the SVI updated with the arbitrage-free interpolation and extrapolation, [Gatheral and Jacquier (2014)] and the latest versions of the quadratic regressions work well in many situations. In our experience, however, we would need another model which combine capabilities of the latter models with better flexibility. For instance, (i) the model should be capable of fitting both smile and skew using the same regression (which could be a problem with the quadratic model); (ii) it would be good to have a separate model parameter which determines location of the smile minimum and could be calibrated to the market data (e.g., in SVI this location is predetermined by the values of the model parameters); (iii) the behavior at wings could be sublinear (see below) while in the original SVI it is strictly linear, etc.

4.3.2   *Parameterization*

From this prospective we do our construction of a new parametric model based on the following assumptions.

(1) As the independent variables of the parametric regression we choose the normalized strike z defined in Eq.(4.1) and time to maturity T.

(2) In the form presented in Eq.(4.6) the model is capable to simulating a different behavior of the smile at call and put wings ([Zhao and Hodges (2013)]). Such a situation could be helpful when modeling commodities where one wing could demonstrate a linear behavior while the other one — sublinear. For the sake of brevity, however, when doing an asymptotic analysis of the model we omit a detailed discussion of sub linearity (to be discussed elsewhere), and concentrate at the case where the variance smile at wings is linear in z.

(3) As there exist multiple justifications that the smile is not symmetric in the z space, it is highly desirable to fit the call and put wings independently.

(4) The parametric function must be continuous in z.

(5) It should be well-behaved close to expiration.

(6) We fit the term structure of the IV term-by-term, i.e., first the variance curve at the first maturity T_1, than the variance curve at the second maturity $T_2 > T_1$, etc. Therefore, we do not consider the dependence of the regression parameters on time. However, we do discuss how to build the whole arbitrage-free IV surface.

(7) As a possible extension of this approach one can rely on the definition of z where the calendar clock T is replaced with a business clock T_v. Here we just mention this opportunity which apparently improves the fitting capability of the model, especially close to expiration, but don't discuss it in detail.

(8) The number of parameters must be minimal.

(9) The parametric function must be fast to evaluate.[5]

(10) The whole IV surface should respect the no-arbitrage conditions.

---

[5]For instance, a recent extension of the SVI model proposed in [Zhao and Hodges (2013)] utilizes the Kummer hypergeometric functions which makes the computation expansive.

Given T, our new parametrization of one term at the IV surface reads

$$w(z) = w_c + \mathcal{S}_C \frac{y}{1+y^2} + F(y)\sqrt{T}\sum_{i=1}^{n} a_i Y^i(y) \qquad (4.6)$$

$$y = z - C, \quad Y(y) \equiv \begin{cases} \frac{1}{\alpha}\mathfrak{S}\left(-\alpha y\right) & y \le 0 \\ \frac{1}{\beta}\mathfrak{S}\left(-\beta y\right), & y > 0 \end{cases},$$

where $w(z)$ is the total implied variance, $w(z) = I^2(z)T$, $I(z)$ is the implied volatility, n determines the maximum degree of the polynomial on $Y(y)$, and $\mathfrak{S}(x)$ belongs to the class of the so-called sigmoid functions, [von Seggern (2007)]. The sigmoid functions tend to some constant at both ends when the argument x tends to $\pm\infty$, and vanish at $x = 0$. Many natural processes, including those of complex system learning curves, exhibit a progression from small beginnings that accelerates and approaches a climax over time. Besides the logistic function, sigmoid functions include the ordinary arctangent, the hyperbolic tangent, the Gudermannian function, and the error function, but also the generalized logistic function and algebraic functions like $x/\sqrt{1+x^2}$.

In Eq.(4.6) the function $F(y)$ defines the model behavior at wings. It could be chosen in such a way that close to $y = 0$ we have $F(y) \propto |y|^{\alpha_0}$ while $F(y) \to y^{\alpha_+}$, $\to \infty$, and $F(y) \to (-y)^{\alpha_-}$, $y \to -\infty$ where $0 < \alpha_- \le 1$, $0 < \alpha_+ \le 1$, $0 < \alpha_0$ are some constants. This construction is accounting for both linear and sublinear behavior of the regression at wings. However, in this paper we will explore only the case $F(y) \equiv |y|$, so the sublinear case will be discussed elsewhere.

From the performance point of view, we want $w(z)$ to be computed with the minimal possible number of computer operations. This guides us in choosing $\mathfrak{S}(x) = \mathrm{erf}(x)$ due to the approximation

$$\mathrm{erf}(x) \approx 1 - (1 + a_1 x + a_2 x^2 + \ldots + a_6 x^6)^{-16},$$

with the maximum error $3 \cdot 10^{-7}$, where $a_1 = 0.0705230784$, $a_2 = 0.0422820123$, $a_3 = 0.0092705272$, $a_4 = 0.0001520143$, $a_5 = 0.0002765672$, $a_6 = 0.0000430638$. This approximation is valid for $x \ge 0$. To use it for the negative x, exploit the fact that $\mathrm{erf}(x)$ is an odd function, so $\mathrm{erf}(x) = -\mathrm{erf}(-x)$ ([Abramowitz and Stegun (1964)]).

It is worth mentioning that using polynomial functions in $\arctan(z)$ was a popular choice among practitioners a while ago, however we don't put this restriction. Also for clarity we fix $n = 2$ and provide a special notation for $\mathcal{S} \equiv a_1$, $\mathcal{K} \equiv a_2$. The reason for this notation will become clear right below.

Under these assumptions, $w(z)$ in Eq.(4.6) has 7 parameters:

- $\mathcal{C}$–shift. This is an edge point between the left and right branches of the smile. For equity options $\mathcal{C} \approx 0$, i.e. this is close to the ATM point. Then, the left branch is a put wing while the right branch is a call wing. For index options the minimum of the smile is usually shifted into positive z. The parameter $\mathcal{C}$ just reflects the value of this shift. Note, that the smile is C_2 at $z = \mathcal{C}$. Indeed, the direct differentiation of $Y(z)$ in the Eq. (4.6) shows that the first derivative is continuous and reads $Y'(z)|_{z=\mathcal{C}} = -1$, while the second derivative vanishes.
- $w_\mathcal{C}$–this is the variance at $z = \mathcal{C}$.
- $\mathcal{S}_\mathcal{C}$–this parameter determines skew of the smile at $z = \mathcal{C}$.
- α–this is a put wing parameter which determines how steep the put wing should be.
- β–this is a call wing parameter which determines how steep the call wing should be.
- $\mathcal{S}$–this parameter determines skew of the smile outside of the region $0 \leq z \leq \mathcal{C}$.
- $\mathcal{K}$–this parameter determines kurtosis of the smile outside of the region $0 \leq z \leq \mathcal{C}$.

In the limit $\alpha \to 0$ or $\beta \to 0$ we obtain $Y(z) \to \mathcal{C} - z$.

Further on, for getting better results we need a minor refinement of the model. As the fitted variance $w(z)$ is expected to be at least C_2 continuous in z, it would be better to eliminate such a non-continuous function as $|z - \mathcal{C}|$. This could be relatively easy done if we find a continuous approximation of the function $|z - \mathcal{C}|$. Among various possible functions we chose that

$$|y| \approx y \tanh[py], \tag{4.7}$$

where p is some constant parameter. Choosing p big enough, say 1000, gives us highly accurate approximation of $|y|$ which is infinitely continuous.

4.3.3   *Asymptotic analysis and meaning of the parameters*

Below we provide an asymptotic analysis of the model to reveal the financial meaning of all the model parameters.

4.3.3.1 *Behavior at $z = C$*

To better understand why one needs a linear correction term, consider the asymptotic behavior of the smile. As $z \to C$ the function $w(z)$ behaves like

$$w(z) \approx w_C + S_C y - \left(S_C + pS\sqrt{T}Y_y(0)\right) y^3 + O(y^4). \qquad (4.8)$$

Thus, this is a polynomial function of $z - C$ which is similar to what [Dumas *et al.* (1998)] model does. More rigorously, it is linear in y at small y if $p < 1/y$, and quadratic if we choose $p \approx 1/y$ at small y. Also as the parameter α determines the steepness of the smile in the put wing, it is reasonable to have an independent parameter to better shape the linear part of the smile near $z = C$. That is why in the Eq. (4.6) we introduced an extra term which is proportional to zS_C at $z \approx C$, and vanishes at $z \to \infty$.

From Eq.(4.8) it is clear that w_C is the total variance at $z = C$, and S_C is the skew at $z = C$, while the kurtosis at $z = C$ vanishes. Also, it is seen that varying p one can change the value of higher moments, which, however, for our analysis is not that important.

Thus, we can interpret the coefficients w_C, S_C and C as some form of adjustment for the critical point not being at $z = 0$.

Note that since the derivatives bear no dependence on β (or α), the model is indefinitely continuous around $z = C$.

4.3.3.2 *Behavior ATM*

Let's consider the behavior of our model at the money when the strike K is equal to the forward price F, and so $z = 0$. To simplify the analysis, we assume $C > 0$, $pC \gg 1$. The value of p could be always chosen such that $pC \gg 1$ unless $C = 0$. Then $\tanh(pC) \approx 1$ if $C > 0$, and $\tanh(pC) \approx -1$ if $C < 0$. For easy of notation, denote $A \equiv Y(-C)$, $A' \equiv Y_y(-C)$, $A'' \equiv Y_{yy}(-C)$.

From the Eq.(4.6) the ATM variance is given by

$$w_0 = w_C - \frac{C}{1+C^2}S_C + AC\sqrt{T}\tanh(pC)\frac{S\alpha + AK}{\alpha^2} + O(y+C). \qquad (4.9)$$

Accordingly, the ATM skew is approximately given by

$$\mathbb{S}_{\text{ATM}} = -\frac{(C^2-1)}{(C^2+1)^2}S_C + \tanh(pC)\sqrt{T}\left[-A(S+KA) + CA'(S+2KA)\right]$$
$$+ O(y+C), \qquad (4.10)$$

and the ATM kurtosis is

$$\mathbb{K}_{\mathrm{ATM}} = -2S_C \frac{C\left(C^2 - 3\right)}{\left(C^2 + 1\right)^3} \tag{4.11}$$
$$+ \sqrt{T}\tanh(pC)\left[-2A'\left(\mathcal{K}(2A - CA') + S\right) + CA''\left(2A\mathcal{K} + S\right)\right]$$
$$+ O(y + C).$$

Further on, we want to determine a connection between the inflection point C and parameters of the smile ATM. In order to do that, first suppose C is small, but our assumption $pC \gg 1$ is still preserved because of a big p. Also, in our numerous experiments where we calibrated this model to various equity and index options with a wide range of maturities and strikes it was observed that a typical value of $\mathcal{K}$ is about 1.0, S is of the order of 1.0, and α varies from 0 to 5. Therefore, from Eq.(4.10) we obtain

$$C^2 = \frac{\mathcal{S}_C - \mathbb{S}_{\mathrm{ATM}}}{3\mathcal{S}_C - 2p\mathcal{S}\sqrt{T}Y_y(0)}. \tag{4.12}$$

Thus, our assumption that the value of C is small is true if $\mathcal{S}_C - \mathbb{S}_{\mathrm{ATM}} \ll 3\mathcal{S}_c - 2p\mathcal{S}\sqrt{T}Y_y(0)$. As C is small (in other words, close to the ATM) the difference $\mathcal{S}_C - \mathbb{S}_{\mathrm{ATM}}$ also has to be small.

At very small T the above solution transforms to

$$\mathbb{S}_{\mathrm{ATM}} = -\frac{\left(C^2 - 1\right)}{\left(C^2 + 1\right)^2} \mathcal{S}_C. \tag{4.13}$$

At small C this equation has the root

$$C = \sqrt{\frac{\mathcal{S}_C - \mathbb{S}_{\mathrm{ATM}}}{3\mathcal{S}_C}}.$$

If C is positive, the ATM point belongs to the put wing, and $\mathbb{S}_{\mathrm{ATM}} < 0$. Therefore, $\mathcal{S}_C$ has to be positive in order for the Eq. (4.12) to be consistent.

Note, that the Eq.(4.12) does not contain α or β, therefore it is valid regardless of whether C is positive or negative. Also from the Eq.(4.10) it follows that the minimum of the smile at $C = 0$ does not coincide with the ATM point.

To illustrate this analysis, here we provide an example of a real smile computed using the proposed model. We run this test on Oct. 7, 2010 and fit the implied volatility of options written on the Eldorado Gold Corporation (EGO) stock with expire on Oct. 15, 2010. The results of fitting are given in Fig. 4.1 where $NS_t \equiv z/\sigma_{ATM}$.

In the below plots and tables, we use $w = I^2(z)$ which is the implied variance, rather than the total implied variance.

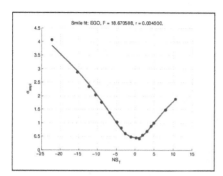 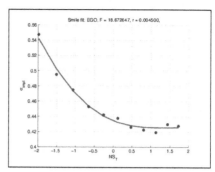

Figure 4.1: Fitting of the IV smile for EGO, $T = 10/15/2010$.

Figure 4.2: Fitting of the IV smile for EGO, $T = 11/19/2010$.

Parameters of the fit found by calibration are given in Tab. 4.1:

Table 4.1: Experiment 1, parameters of the fit.

w_C	$\mathcal{S}_C$	$\mathcal{C}$	$\mathcal{S}$	$\mathcal{K}$	α	β
0.1652	-0.04302	0	-0.20	1.035	-0.42623	0.60308

Thus, this smile does not demonstrate any shift of the minimum from ATM.

In the second example we fitted the next term of the same product. The results are given in Fig. 4.2. Parameters of the fit found by calibration are given in Tab. 4.2:

Table 4.2: Experiment 2, parameters of the fit.

w_C	$\mathcal{S}_C$	$\mathcal{C}$	$\mathcal{S}$	$\mathcal{K}$	α	β
0.16775	0	0.5769	-0.003	0.11	-0.0004	2.7457

It is seen that in the test $\mathcal{C}$ is not a small parameter, therefore simple approximations suggested in the above cannot be used in this case.

The third example is given for options written on Financial Select Sector SPDR Fund (XLF) stock, also for the front term Oct.15, 2010. The results are given in Fig. 4.3.

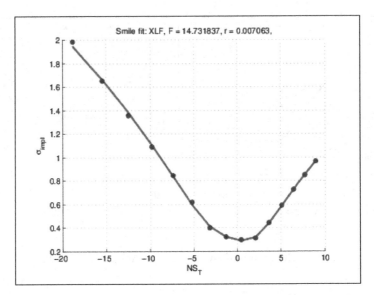

Figure 4.3: Fitting of the IV smile for EGO, $T = 10/15/2010$.

Parameters of the fit found by calibration are given in Tab. 4.3:

Table 4.3: Experiment 3, parameters of the fit.

w_C	S_C	C	S	K	α	β
0.0703	−0.038	0.0032	−0.2	0.99	0.5741	−0.80175

In this test the calibrated value of C is small, so one can use the proposed approximations which connect the ATM skew and kurtosis with the value of C.

4.3.3.3 *Behavior At Infinity*

As $K \to \infty$ at fixed S, so does z. Assume, for example, that $\mathfrak{S}(x) \equiv \arctan(x)$. Expanding variance in series around positive infinity, we have

$$w(z) \approx w_C + \frac{\sqrt{T}}{\beta^3}\left(\beta S - \pi K\right) + \pi\sqrt{T}\frac{\frac{1}{2}\pi K - S\beta}{2\beta^2}y + O\left(1/y\right). \quad (4.14)$$

Hence, the variance is linear in log-moneyness χ at positive infinity, with the slope

$$\phi(\infty) = \frac{1}{2}\pi\beta^{-2}(\frac{1}{2}\pi K - S\beta). \quad (4.15)$$

This well agrees with the result of [Lee (2004)]. Thus, our interpretation of β follows: this parameter controls the slope of the smile at the infinite strike.

At $K \to 0$, $z \to -\infty$. Expanding variance in series around negative infinity, we have

$$w(x) \approx w_C - \frac{\sqrt{T}}{\alpha^2}(\alpha\mathcal{S} + \pi\mathcal{K}) - \pi\sqrt{T}\frac{\frac{1}{2}\pi\mathcal{K} + \mathcal{S}\alpha}{2\alpha^2}y + O\left(1/y\right). \quad (4.16)$$

Hence, the variance is also linear in log-moneyness χ at negative infinity, with the slope

$$\phi(-\infty) = -\frac{1}{2}\pi\alpha^{-2}(\frac{1}{2}\pi\mathcal{K} + \mathcal{S}\alpha). \quad (4.17)$$

This also agrees with the result by [Lee (2004)]. Accordingly, our interpretation of α is: this parameter controls the slope at strike close to zero.

Close to expiration z tends to infinity. However, for our function in the Eq.(4.6) this is not a problem. Indeed, at $T \to 0, z \to \infty$, the product $z\sqrt{T} \to \log K/F$, therefore from Eq.(4.14)

$$w(z) \to w_C + \frac{\pi}{2\beta^2}\left(\frac{1}{2}\mathcal{K}\pi - \beta\mathcal{S}\right)\log\frac{K}{F}, \quad (4.18)$$

As mentioned in [Medvedev and Scaillet (2008); Ledoit *et al.* (2002)] in diffusion models, the ATM implied volatility is known to converge to the spot volatility when T goes to zero. In our case from Eq.(4.18) the ATM value $w(z)|_{z=0} = w_C$, which implies $I(z) = \sigma_C = const$, $w_C = \sigma_C^2 T$. This provides another interpretation of the parameter σ_C as the IV of the underlying stock at $T = 0$.

4.3.4 *No arbitrage conditions*

In Chapter 3 the local volatility (or IV) surface is built based on the stochastic mode which guarantees no-arbitrage by construction. Here our regression-based approach doesn't provide such a nice feature *per se*. Therefore, a special care should be taken under calibration in order not to introduce arbitrage into the IV surface. See, for instance, [Andreasen and Huge (2011); Lipton and Sepp (2011b); Gatheral and Jacquier (2014)] and references therein.

The no-arbitrage conditions could be expressed in various forms. One of the approaches is to say that the local volatility function must be non-negative. The reason for that is that the local volatility function is directly related to the pdf (density) of the underlying, which in turn has to be non-negative. For convenience, here we recall Eq.(3.2) which expresses the local volatility via $w(z)$

$$\sigma_{loc}^2(T,K) = \frac{\partial_T w}{\left(1 - \frac{x \partial_x w}{2w}\right)^2 - \frac{(\partial_x w)^2}{4}\left(\frac{1}{w} + \frac{1}{4}\right) + \frac{\partial_x^2 w}{2}}, \qquad (4.19)$$

The nominator of this expression is the so-called calendar spread, and the denominator of it is equivalent to the so-called butterfly spread (which for the call option with price $C(T,K)$ is defined as $\frac{\partial^2 C(T,K)}{\partial K^2}$, [Gatheral and Jacquier (2014)]). Both spreads must be non-negative for no-arbitrage.

However, as shown in [Carr and Madan (2005)], one more condition is required in addition to the above mentioned, which tells that so-called vertical call spread (which for the call option with price $C(T,K)$ is defined as $\frac{\partial C(T,K)}{\partial K}$) should be negative for the call options, or the vertical put spread should be positive for the put options. For the IV these conditions for the vertical spreads could be transformed to the following, [Carr (2004)]

$$\frac{R(d_2)}{\sqrt{T}} \leq K \frac{\partial I(K,T)}{\partial K} \leq \frac{R(-d_2)}{\sqrt{T}}, \qquad (4.20)$$

where $R(d) \equiv \frac{1 - N(d)}{N'(d)}$ is Mill's ratio, $N(d)$ is the normal cdf, and d_2 comes from the Black-Scholes formula. The convenience of such a representation lies in the fact that Mill's ratio for the standard normal distribution reads

$$R(x) = e^{x^2/2}\sqrt{\frac{\pi}{2}}\,\mathrm{erfc}\left(\frac{x}{\sqrt{2}}\right).$$

The latter can be efficiently computed by the particularly simple continued fraction representation at $x > 1$

$$R(x) = \cfrac{1}{x + \cfrac{1}{x + \cfrac{2}{x + ...}}}, \qquad (4.21)$$

or by using Taylor series expansion at $0 \leq x \leq 1$, [Gasull and Utzet (2014)]. The IV surface should also satisfy the asymptotic conditions discussed in the previous section, namely: the slope $\phi(\infty)$ of the call wing at $z \to \infty$ should be $0 \leq \phi(\infty) \leq 2$, and the slope of the put wing $\phi(-\infty)$ at $z \to -\infty$ should be $0 \geq \phi(-\infty) \geq -2$.

Being equipped with all these no-arbitrage conditions, the next step to consider is the construction of the IV surface in the domain (T,K). This is not a problem if, say, we want to have the IV surface to be defined at some discrete grid in the (T, K) space (that could be a grid where we want the local volatility function to be determined — a standard approach when one calibrates the LSV model to the market data): $\mathbb{G} : [T_i \times K_j]$, $i \in [1, N]$, $j \in [1, M]$ under two assumptions made: (i) for every grid node (i, j) there exists a market quote $Q(T_i, K_j)$ which is an option price (call or put, or both); (ii) there is no need to ever know the IV at other possible values of T, K which don't belong to $\mathbb{G}$. Certainly, in practice both assumptions are unrealistic. Therefore, some kind of interpolation/extrapolation which preserves no-arbitrage is necessary.

Therefore, in order to calibrate our model to the market data such that not every node on the computational grid is provided with a corresponding market quote, a special calibration algorithm ws elaborated on which in more detail is described in the next Section.

4.3.5 *Finding parameters for one term*

To obtain the values of the smile parameters, a non-linear least square optimization is used. Every market point is taken with some weight which is usually of the following form

$$w(z) = \frac{1}{2} \left(w_c(z) + w_p(z) \right), \tag{4.22}$$

$$w_c(z) = (1 - |\Delta_c|) \min \left[0.1, \left(\frac{z}{\sigma_{atm}} \right)^\nu \right]$$

$$w_p(z) = (1 - |\Delta_p|) \min \left[0.1, \left(\frac{z}{\sigma_{atm}} \right)^\nu \right].$$

Here Δ_c and Δ_p are the market call and put deltas of the option, σ_{atm} is the ATM market implied volatility, ν is some parameter which is typically taken to be -2 or -3. Having these weights, the following optimization problem was solved to obtain parameters of the fit

$$\min_{p_1 \ldots p_7} \sum_{i=1}^{N} W_i(z) \left[w_m(z_i) - w(z_i, p_1 \ldots p_7) \right]^2, \tag{4.23}$$

where N is the total number of the raw option data, $W_i(z)$ is the weight of the ith point, w_m is the market total implied variance of the data, $\nu_i, i = 1, 7$ are the parameters of the model.

This minimization problem is solved under the whole bunch of no-arbitrage constraints discussed in the previous section. The no-arbitrage constraints are checked at every node on the grid $\mathbb{G}$, while the asymptotic slope is checked at two edge points on the grid for every time slice.

Further on, we calibrate all terms, provided as an input, that contain at least a single data point, by using bootstrap, i.e. term by term. We start with reordering all the market data in the ascending order, and then proceed with fitting the shortest term at $T = T_1$. Then the next term at $T = T_2$ is fitted, etc. To solve this optimization problem we use a genetic algorithm implemented in [Hansen (2008)] which we updated with allowance for the equality and inequality constraints. This algorithm guarantees finding a global minimum. A typical time necessary to get the values of the parameters for one term in C++ is about 0.5 secs with the maximum number of function evaluations set to 10^4. Based on our experiments this value provides a very good fit, while it could be lowered to get a better performance. Note that as our algorithm belongs to the class of evolutionary optimization ([Simon (2013)]) it is very suitable for parallelization.

We also need to underline that the above minimization problem is solved at points z_i, $i = 1...s$ where for the given term T_j the market prices are available at s strikes $\hat{K}_{i,1} = 1...s$ such that $z_i = (\log \hat{K}_i / F(T_j)) / \sqrt{T_j}$, $i = 1...s$. However, the no-arbitrage constraints are checked at another set of points: that ones that belong to the $\mathbb{G}$ grid. According to the definition of $\mathbb{G}$ these points are $z_i = (\log(K_i)/F(T_j))/\sqrt{T_j}$, $i = 1...M$. By construction the implied volatilities obtained on the grid nodes are arbitrage-free.

Smart initial guess. When calibrating every term at the beginning we use a special algorithm to provide a good initial guess. The idea behind this algorithm is that as follows from Eq.(4.6) $w_{zz}(z) = 0$ at the point $z = \mathcal{C}$. Therefore, one can look at the input IVs in the z space, compute the second derivative and find where it vanishes. In case the second derivative is positive everywhere, as the value of $\mathcal{C}$ one can take such z where the IV is minimal among all values belonging to this term. This construction also works well when the IV surface has a skew, not a smile, which is typical for index options.

Given $\mathcal{C}$ other parameters could be obtained relatively straightforward. Indeed, from Eq.(4.6)

$$w_C = w(z)|_{z=C}, \qquad \mathcal{S}_C = w_z(z)|_{z=C}. \tag{4.24}$$

Now denote

$$\kappa(y) = \frac{1}{y \tanh(py)\sqrt{T}} \left(w(y + \mathcal{C}) - w(\mathcal{C}) - \mathcal{S}_C \frac{y}{1 + y^2} \right), \qquad y \equiv z - \mathcal{C},$$

so based on Eq.(4.24) $\kappa(y)$ is a known function of y. Accordingly Eq.(4.6) could be re-written in the form

$$\mathcal{S}Y(y) + \mathcal{K}Y^2(y) = \kappa(y). \tag{4.25}$$

To find the initial guess for the remaining parameters $\alpha, \beta, \mathcal{S}, \mathcal{K}$ we need 4 additional market IVs. At least two of them should lie on the different sides of the IV curve with regard to the point $y = 0$. As an example, consider three points $y_1 > y_2 > y_3 > 0$. Then, $Y(y)$ in Eq.(4.25) is defined via parameter β, see Eq.(4.6). Using Eq.(4.25) with y_1 and y_2 we find

$$\mathcal{S} = \frac{\kappa(y_1)Y^2(y_2) - \kappa(y_2)Y^2(y_1)}{Y(y_1)Y^2(y_2) - Y(y_2)Y^2(y_1)}, \qquad \mathcal{K} = \frac{\kappa(y_2)Y(y_1) - \kappa(y_1)Y(y_2)}{Y(y_1)Y^2(y_2) - Y(y_2)Y^2(y_1)}. \tag{4.26}$$

Now using the point y_3 we numerically solve the equation Eq.(4.25) with regard to β.

Since parameters $\mathcal{S}, \mathcal{K}$ are already found, the last point $y_4 < 0$ could be used together with Eq.(4.25) to numerically solve for α. This finalizes computation of the initial guess.

In case the input data points are located as $y_1 > y_2 > 0 > y_3 > y_4$ the easiest way is to add an extra point to the negative y by using interpolation, and then remove one point, e.g., y_2 from the positive y, thus getting back to the previous case.

4.3.6   *No-arbitrage interpolation on the grid*

Various approaches were discussed in the literature with regard to this problem. For instance, an arbitrage-free interpolation was considered in [Andreasen and Huge (2011); Fengler (2005); Gatheral and Jacquier (2014)] (see also references therein). Here, however, we suggest another approach, which is similar in spirit to that in [Gatheral and Jacquier (2014)].

If one works in the z space given T the usual approach would be to choose some number γ such that all $z = [z_1...z_M]$ for this term are in the range $-\gamma < z/\sigma_* < \gamma$. Here σ_* is some normalization constant which doesn't depend on T. By financial meaning, σ_* could be chosen as the ATM IV which corresponds to the shortest maturity, This is the most liquid strike of the instrument, and usually it is pretty well-known from the market data. In other words, for the IV surface just a range of γ standard deviations in both up and down directions from the ATM is taken into account. Outside of this domain the remaining strikes are treated to be illiquid, and, therefore, they are taken out of consideration. In practical applications $\gamma = 5$ could be chosen, but this assumption could be easily relaxed.

Another situation is if we want the IV surface to be a building block of the numerical method which solves the pricing/calibration problem using the local stochastic volatility model. The idea is first to calibrate the IV surface to the market quotes of the vanilla options, and then compute the local volatility surface using the Dupire's formula. In this case, we need the values of the local volatility function not only at strikes and maturities available were the market data are available, but at all nodes in the computational domain K, T which is these calculations. In more detail, in this case we fist define a fixed domain in K space: $[K_1...K_M]$. Accordingly, for the z variable we have a map $z_i = \log(K_i/F(T_j))/\sqrt{(T_j)}$ which depends on the current expiration T_j, $j = 1...N$. In other words, we work on the $\mathbb{G}$ grid which was described in the above.

Provided by a set of the IV market data for expirations $T_1 < T_2 < ...$ $< T_m$ (these expirations in general don't coincide with the temporal nodes of the grid $\mathbb{G}$, but could be a subset of that) we calibrate our model term by term based on the algorithm of Appendix 4.3.5. To remind, this algorithm takes into account the entire set of the no-arbitrage constraints at every point on the given grid $\mathbb{G}$. By construction, despite the market provides the option quotes per strikes, the grid was built in the z space, not in the strike K space. At the end we obtain all values $w(z, t)$ where $z \in [-\gamma\sigma_* = z_1, ..., z_N = \gamma\sigma_*], T \in [T_1, ..., T_m]$. Given thus found $w(z, t)$ the corresponding undiscounted call and put prices can be further obtained by using the Black-Scholes formula afterwards.

After this step is completed the arbitrage-free values $w(z, t)$ become available for the terms with expirations $T_1 < T_2 < ... < T_m$. For the sake of clearness let us denote them as $w_1(z, t) \equiv w(z, t)$, $t \in [T_1, ..., T_m]$. However, our grid $\mathbb{G}$ by construction also might contain some other expirations $\hat{T}_1, ..., \hat{T}_l$ where l is the total number of temporal nodes on the grid. Let us denote this set as $w_2(z, t)$. Also let us emphasize that the space nodes z are the same for both $w_1(x, t)$ and $w_2(t)$ by construction.

Therefore, to find $w_2(z, t)$ next we need to interpolate $w_1(x, t)$ to the expirations of $w_2(z, t)$ at every point z on the $\mathbb{G}$ grid. When doing that it is more convenient to proceed in the pricing space despite this is a bit more computationally intensive as we need to convert the IVs force to the prices at the beginning of this step, and back to the IVs at the end of this step. For the fitted terms after calibration is done we already know all parameters of the fit, so we are able to compute the call option value at any point at the $\mathbb{G}$ grid. And the no-arbitrage conditions were already respected in these points as well. Also we know the corresponding map $K_i \to F(T_j)e^{z_i\sqrt{T_j}}, i = 1...M, j = 1...N$.

Accordingly, in K space the no-arbitrage conditions for the call option: non-negativity of the calendar and butterfly spreads and non-positivity of the vertical spread read

$$\frac{\partial C(K,T)}{\partial T} \geq 0, \qquad \frac{\partial C(K,T)}{\partial K} \leq 0, \qquad \frac{\partial^2 C(K,T)}{\partial K^2} \geq 0. \qquad (4.27)$$

Now chose a monotonic time interpolation of $C(K,T)$ at K=const of the form

$$C(K,T) = \alpha(T)C(K,T_1) + [1 - \alpha(T)]C(K,T_2) \qquad (4.28)$$

where $T_1 < T < T_2$ and

$$\alpha(T) = \frac{a(T_2) - a(T)}{a(T_2) - a(T_1)}, \qquad (4.29)$$

where $a(T)$ is some monotonic function. Obviously, $\alpha(T) \in [0,1]$, and $\alpha(T)$ doesn't depend on K. And this is a valid interpolation formula in a sense that the values of $C(K,T)$ at $T = T_1$ and $T = T_2$ coincide with $C(K,T_1)$ and $C(K,T_2)$. Also thus defined $C(K,T)$ provides

$$\frac{\partial C(K,T)}{\partial T} = \alpha(T)\frac{\partial C(K,T_1)}{\partial T} + [1 - \alpha(T)]\frac{\partial C(K,T_2)}{\partial T} \qquad (4.30)$$
$$+ \frac{\partial \alpha(T)}{\partial T}[C(K,T_1) - C(K,T_2)] \geq 0,$$

if $\partial_T\alpha(T) < 0$, i.e., $\partial_T a(T) < 0$. That is because we constructed $C(K,T_1), C(K,T_2)$ such that they obey the no-arbitrage condition $\partial_T C(K,T)|_{T=T_i} > 0$, $i = 1,2$.

It is easy to see that Eq.(4.28) also solves the second and third lines in Eq.(4.27) provided that these conditions were met at $T = T_1$ and $T = T_2$. The latter follows from our construction at the previous step of the algorithm. Also it can be shown that this expression still preserves the extreme slopes of the interpolated terms (that are at $z \to \infty$ and at $z \to -\infty$) to follow the asymptotic conditions provided by [Lee (2004)]. To see that, note that the latter could be represented in the form $C(K,T,I) < C(K,T,\sqrt{2|\chi|/T})$. Therefore, our interpolation provides

$$C(K,T,I) = \alpha(T)C(K,T_1,I_1(K,T_1)) + (1 - \alpha(T)C(K,T_2,I(K,T_2))$$
$$< \alpha(T)C(K,T_1,\sqrt{2|\chi_1|/T_1}) + (1 - \alpha(T)C(K,T_2,\sqrt{2|\chi_2|/T_2})$$
$$< C(K,T,\sqrt{2|\chi|/T}). \qquad (4.31)$$

The last equality holds because we interpolate at K=const, S=const, so $C(K, T, \sqrt{2|\chi|/T})$ is a function of T only, and this is a concave function of T.

As far as extrapolation is concerned, in addition to the no-arbitrage conditions we need to prove that the extreme slopes of the extrapolated terms (that are at $z \to \infty$ and at $z \to -\infty$) still preserve the asymptotic conditions provided by [Lee (2004)].

We show that an extrapolation formula

$$T^k C(K, T) = \alpha(T) T_1^k C(K, T_1) + [1 - \alpha(T)] T_2^k C(K, T_2), \qquad (4.32)$$

with $k \in Re$, $k \leq -0.5$ is suitable for this purpose.

Indeed, similar to Eq.(4.28)

$$T^k C(K, T, I) = \alpha(T) T_1^k C(K, T_1, I_1(K, T_1)) + (1 - \alpha(T) T_2^k C(K, T_2, I(K, T_2))$$
$$< \alpha(T) T_1^k C\left(K, T_1, \sqrt{2|\chi_1|/T_1}\right) + (1 - \alpha(T) T_2^k C\left(K, T_2, \sqrt{2|\chi_2|/T_2}\right)$$
$$< T^k C(K, T, \sqrt{2|\chi|/T}). \qquad (4.33)$$

The last inequality holds because we interpolate at K=const, S=const, so

$$f(T) \equiv T^k C(K, T, \sqrt{2|\chi|/T}), \qquad (4.34)$$

is a function of T only, and $k < 0$ is chosen such that $f(T)$ is a convex function. The latter condition depends on the value of the interest rate r, and χ. Usually, $k = -1$ is sufficient even for the ATM strikes.[6] Also thus defined $C(K, T)$ provides

$$T^k \frac{\partial C(K, T)}{\partial T} = \alpha(T) T_1^k \frac{\partial C(K, T_1)}{\partial T} + [1 - \alpha(T)] T_2^k \frac{\partial C(K, T_2)}{\partial T} \qquad (4.35)$$
$$+ \frac{\partial \alpha(T)}{\partial T} [T_1^k C(K, T_1) - T_2^k C(K, T_2)] - k T^{k-1} C(K, T) \geq 0,$$

if $\partial_T \alpha(T) < 0$. That is because we constructed $C(K, T_1), C(K, T_2)$ such that they obey the no-arbitrage condition $\partial_T C(K, T)|_{T=T_i} > 0$, $i = 1, 2$, and $T_1^k C(K, T_1) - T_2^k C(K, T_2) < 0$ since $T_1 < T_2$ and $k = -1$. Thus, the calendar spread is non-negative for the call option. The other two no-arbitrage conditions obviously follow.

---

[6] As it could be easily seen from analysis of the Black-Scholes formula for the call option prices this is the most sensitive region. Therefore, the choice of, e.g., $k = -0.5$ could make $f(T)$ to be concave close to ATM.

4.3.7 Numerical experiments

4.3.7.1 Stability of the fitted parameters

In a typical experiment a volatility smile of an option written on the S&P 500 index (SPX) was fitted using the proposed model. The raw data S&P are collected for $T = 0.6247$ (228 days to expiration), $F = 76.58, T_v = 0.6215$. We find that our model fits the data pretty well, with parameters of the fit obtained by running the above described minimization algorithm, given in Tab. 4.4.

Table 4.4: Values of the parameters obtained in the test

w_C	$\mathcal{S}$	$\mathcal{K}$	α	β	$\mathcal{C}$	$\mathcal{S}_C$
0.0435	−0.763	74.5	3.12468	1.5000	0.2739	−0.05921

It is interesting to see, however, how the fitting parameters behave as a function of time. In other words, what is the sensitivity of the fit to changes in time. To investigate this we use the same data on the SPX closing IVs given for 133 sequential days and plot time-series of the model parameter values. These results are given in Figs. 4.4–4.10.

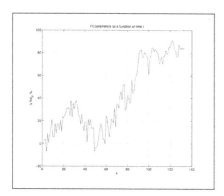

Figure 4.4: Sensitivity of w_C to the time change.

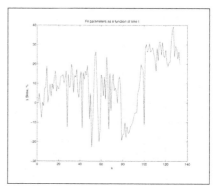

Figure 4.5: Sensitivity of $\mathcal{S}$ to the time change.

As one can see, the most time sensible parameters are α and β. So they need to be refitted more often, probably few times a day. At the time scale of a few days other parameters change just within 10–20 %; therefore, they could be refitted less often.

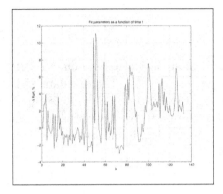

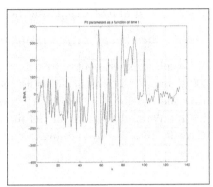

Figure 4.6: Sensitivity of $\mathcal{K}$ to the time change.

Figure 4.7: Sensitivity of $\mathcal{C}$ to the time change.

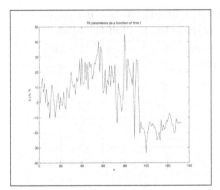

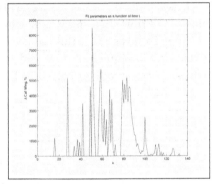

Figure 4.8: Sensitivity of $\mathcal{S}_C$ to the time change.

Figure 4.9: Sensitivity of β to the time change.

We want to emphasize that by construction our model provides just the static fit of the current market snapshot of the options IVs, and does not consider any dynamics of the IV surface. Therefore, the dependence of the model parameters on time serves just to the illustrative purposes and helps in organizing a rapid calibration procedure. This does not mean that looking at the time dependence of the model parameters one can make a predictive conclusion on how the future IV behaves with time.

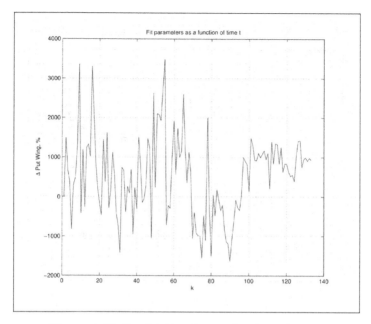

Figure 4.10: Sensitivity of α to the time change.

4.3.8 *Constructing a local volatility surface*

In this example we take data from http://www.optionseducation.org on
XLF traded at NYSEArca on March 25, 2014. The spot price of the index
is $S = 22.64$, the interest rate $r = 0.0148$. The option IVs are given in
Tab. 4.5. We take all OTM quotes and some ITM quotes which are very
close to the ATM. At the overlapped strikes for calls and puts we take an
average of I_{call} and I_{put} with weights proportional to $1 - |\Delta|_c$ and $1 - |\Delta|_p$
correspondingly.[7] We use the proposed parametric fit to construct the IV
surface at all given expirations and strikes in a range $K \in [17, 28]$ with the
step 0.5. In doing so we calibrate the first and the last term using the above
described algorithm. The other terms are found on the grid by applying the
arbitrage-free interpolation with $a(T) = C(K, T, I(K, T))$ and $K = 17$ (in
this case $I(K, T)$ is assumed to be provided in the given set of data).

---

[7]By doing so we do take into account effects reported in [Ahoniemi (2009)] that the IVs
calculated from identical call and put options have often been empirically found to differ,
although they should be equal in theory. However, our weights are a pure empirical rule
of thumb, and more detailed investigation of this is required.

Table 4.5: XLF option IVs: C–call options, P–put options.

T	K, Put					
	18	19	20	21	22	23
4/19/2014	–	32.90	26.79	20.14	15.19	12.93
5/17/2014	33.27	26.88	23.08	18.94	16.12	13.86
6/21/2014	27.84	23.90	21.07	18.88	16.95	15.82
7/19/2014	26.09	22.81	20.29	18.13	16.30	14.93
9/20/2014	24.20	22.23	20.32	18.76	17.40	16.41
12/20/2014	23.75	22.09	20.67	19.44	18.36	17.60

T	K, Call							
	21	22	23	24	25	26	27	28
4/19/2014	–	15.79	13.38	15.39	–	–	–	–
5/17/2014	16.71	14.48	–	13.75	–	–	–	–
6/21/2014	16.31	14.78	–	13.92	14.28	16.58	–	–
7/19/2014	16.82	15.24	–	14.36	14.19	15.20	–	–
9/20/2014	17.02	15.84	–	14.99	14.56	14.47	14.97	16.31
12/20/2014	17.63	16.61	–	15.86	15.47	15.12	15.18	15.03

The results of this fitting are given term-by-term in Fig. 4.11. Accordingly, thus constructed the IV surface is represented in Fig. 4.12, and the local volatility surface and the implied density obtained from the IV surface by applying the Dupire's and Breeden-Litzenberger formulas are given in Figs. 4.13–4.14.

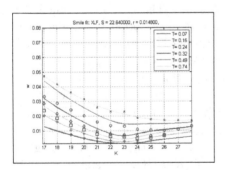

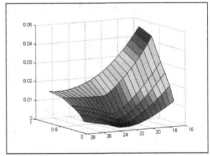

Figure 4.11: Term-by-term fitting of the IV surface constructed using the whole set of data in Tab. 4.5.

Figure 4.12: The IV surface obtained using the first and last terms in Tab. 4.5 and interpolation.

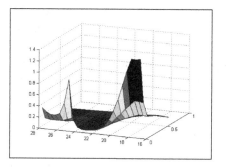
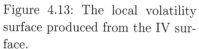
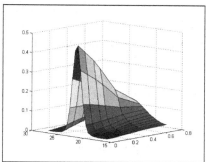

Figure 4.13: The local volatility surface produced from the IV surface.

Figure 4.14: The density implied from the IV surface in Fig. 4.12.

One can see that the local volatility is positive everywhere on the grid which is provided by (i) using the no-arbitrage constraints when calibrating each term, and (ii) using the arbitrage-free interpolation instead of calibration for some terms.

Also the above results clearly show that calibration provides a very good fit to the market data for the first and the last term. However, for the other terms the no-arbitrage conditions could be very restrictive. Therefore, the genetic algorithm requires many evaluations of the objective function, and could be slow. In contrast, the arbitrage-free interpolation is very fast but doesn't give such a good fit to the market data.

4.4 Discussion

When fitting the IV surface we rely on raw quotes for liquid options provided by the market. Unfortunately, markets are different. For instance, in the oil market only a few strikes are traded as European listed options, usually the ATM, one ITM and one OTM option. Other strikes are traded OTC and, moreover, as Asian options. Also, the smile behavior at wings could be different for index and equity options. The variance smile could still be linear in z at wings but with a very different skew. And it seems there is no a clear theoretical reason why it could not be. Therefore, if somebody has just an intuition on how the smile wings should behave, he/she could better rely on this intuition rather than on some unreliable illiquid data, and treat the latter as outliers.

Fortunately, the proposed model is able to address such an intuition by doing the following trick. Suppose that we want to have the new model for the index options smile at the call wing being as close as possible to the existing smile produced by some proven (reference) model. Then we can move the value of the call wing parameter β into a different region by imposing a special constraint. By doing that, we make the fit a bit worse, but thus found slope (after the minimization is done) turns to be closer to the corresponding reference model skew. And we still preserve the continuity of the model.

Another issue with the model is as follows. Suppose, for a given term the number of strikes for which the market quotes are available, is less than the number of the model parameters, i.e. 7. In this case the parameterization is over determined. The no-arbitrage constraints and the asymptotic behavior of the smile help to resolve this however could not be sufficient. For instance, if only the ATM quote is liquid and available. In this situation we have either to reduce the number of parameters, or to use some tricks.

To give an example of such a trick consider the case when only a single quote I corresponding to the strike K is available given the time to expiration T. Under this situation it doesn't make sense to calibrate our parameterization to this single point. Instead, we treat the entire term as fully unknown, and find the IV values by using the arbitrage-free interpolation in time as it was described in the above. However, to fit exactly thus found IVs to the given quote we exploit the remaining flexibility in the definition of the function $a(T)$. We remind that $a(T)$ was not yet defined explicitly, and only indirectly via the condition $\partial_T a(T) > 0$. Then taking $a(T) = C(K, T, I)$ provides this inequality on one hand. On the other hand, as it could be easily checked, thus defined $a(T)$ matches exactly the given quote $C(K, T, I)$.

When 3 or 4 quotes are available for the given term, the kurtosis $\mathcal{K}$ is a natural candidate to be removed from the parameterization. In other words, we fix $\mathcal{K} = 0$ and so reduce the total number of parameters to 6. Also, $\mathcal{S}_C$ could be the next preferable choice to omit.

As far as the relation of the proposed model to the existing ones and comparison of our results with that obtained using, e.g., the SVI model we want to underline the following. There are at least two problems that, e.g., quantitative analysts are dealing with pretty often, and that require knowledge of the implied volatility. One, which is important for traders and market makers, is to fit the existing market IV data, basically on a term-by-term basis, so the IV values in between of the tradable strikes could be out of their interest. This problem could be solved sufficiently well using various popular models, including SVI and the modern quadratic fit, while the proposed in this paper model also falls into this class. Then to choose an appropriate model questions about stability of the model parameters, uniqueness of the set of parameters that provide a reasonable fir to the given smile/skew should be addressed. An interesting discussion on this subject as applied to the SVI model could be found in ([Nuclear Phynance (2007)]). The SVI model often produces a non-unique set of the calibrated parameters in a sense that using various initial guess es in the calibration procedure one can get different sets of the SVI parameters that fit the given market data with almost same accuracy. This means that stability of fitting parameters could be in question. In our model to eliminate a possible instability a practical recipe is as follows. We calibrate the model for the first time, and then, when after some period of time we need to refit it, we fix one of the parameters, for instance, C, at the previous level, thus refitting only the remaining parameters. Then in our experience the daily variations of the model parameters are pretty much suitable for traders, i.e., the fit could be treated as stable.

The second and more challenged problem is to build a local volatility function for some grid pricer given the market option quotes. This could be addressed by first building the IV surface and then using the Dupire's formula. Here a good quality of the fit at the given quotes is not sufficient, and in addition no-arbitrage constraints in every grid point in the (K, T) space as well as the correct asymptotic behavior of the smile/skew should be preserved. Under these restrictive conditions perhaps any model, including SVI, which operates just with 5 model parameters is not flexible enough to be able to meet all the constraints. Thus, it either sacrifices by the quality of the fit, or by the no-arbitrage conditions, or by the correct asymptotic behavior in order to converge. Finally, finding the correct fit could be slow because of these limitations. As a possible resolution in this paper we demonstrate that fitting just the first and the last term and then using the arbitrage-free interpolation could be a reasonable alternative from both performance and goodness of the fit point of view. Also as our model contains more parameters, it provides an additional flexibility to better solve the above constrained optimization problem.

To illustrate this in Figs. 4.15–4.18 we present the results of the test described in Section 4.3.8, where now instead of our model the SVI model was used. When calibrating this model in the first test the no-arbitrage constraints on a grid were not taken into account except positivity of the total variance w.

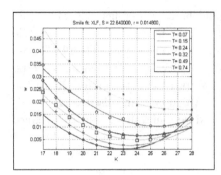
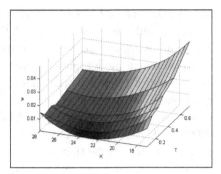

Figure 4.15: Term-by-term fitting of the IV surface based on the data in Tab. 4.5 and the SVI model.

Figure 4.16: The local volatility surface produced from the IV surface using the SVI model.

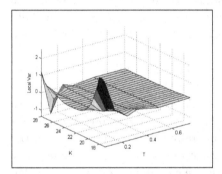
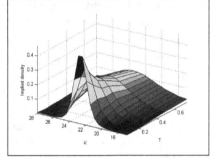

Figure 4.17: The local volatility surface built from the SVI IV surface.

Figure 4.18: The density implied from the SVI IV surface shown in Fig. 4.16.

The quality of the IV fit is good, however, the no-arbitrage conditions in the strike space are not validated as well as the calendar arbitrage could be observed for the OTM call options. Accordingly, the grid local volatility is negative at some strikes and expirations. However, the implied density is positive on the grid.

Surprisingly, if we take the no-arbitrage constraints into account and repeat the above test, at least in our numerical experiments the optimizer was never able to find a good fit. Moreover, thus found model parameters always produced negative local volatilities at some strikes and expirations. This justifies our hypothesis that constructing the local volatility surface on a grid by using the implied volatility surface calibrated to the market data with the SVI model could be inefficient. Most likely bad fit (and as the consequence — negative local volatilities) are observed due to insufficient flexibility of the SVI model which has only 5 parameters per one expiration term.

PART 3
The Variance Gamma Model

Chapter 5

An Expanded Local Variance Gamma Model

In this part of the book we consider another local volatility model. In this model the underlying is driven by a Variance Gamma process of [Madan and Seneta (1990)], rather than the Geometric Brownian Motion, but also equipped with a local volatility function. Such a model was first proposed in [Carr and Nadtochiy (2014)] to (i) improve computational efficiency of calibration of the local volatility surface, and (ii) to built a richer flavor of the local volatility model. The latter is achieved by adding a stochastic volatility component via a stochastic change of time. We will discuss this in more detail in what follows.

The original paper [Carr and Nadtochiy (2014)] was then re-worked to [Carr and Nadtochiy (2017)] where the Local Variance Gamma (LVG) model has been fully elaborated as well as calibration of the model was provided. With a bit more detail, the model assumes that the risk-neutral process for the underlying futures price is a pure jump Markov martingale, and that European option prices are given at a continuum of strikes and at one or more maturities. The authors construct a time-homogeneous process which meets a single smile and a piecewise time-homogeneous process, which can meet multiple smiles. However, in contrast to eg, [Itkin and Lipton (2018)], their construction leads not to a PDE, but to a partial differential difference equation (PDDE), which permits both explicit calibration and fast numerical valuation. In particular, it does not require application of any optimization methods, rather just a root solver. In [Carr and Nadtochiy (2017)] this model is used to calibrate the local volatility surface assuming its piecewise constant structure in the strike space.

One of the potential criticism of this calibration method is the fact that the resulting local volatility function has a finite number of discontinuities. So it would be advantaged to relax the piecewise constant behavior of the surface. This is similar to how [Itkin and Lipton (2018)] was developed to overcome the same problem as compared with [Lipton and Sepp (2011a)].

On this way, recently [Falck and Deryabin (2017)] applied the LVG model to the FX options market where usually option prices are quoted only at five strikes. They assumed that the local volatility function is continuous, piecewise linear in the four inner strike subintervals and constant in the outer subintervals. A closed form solution of the PDDE derived in [Carr and Nadtochiy (2014)]) is obtained with this parametrization, and calibration of some volatility smiles is provided. Still, to calibrate the model the authors rely on a residual minimization by using a least-square approach. So, despite an improved version of the LVG model is used, computational efficiency of this method is not perfect.

Another remark of [Carr and Nadtochiy (2017)] is about the limitation that the risk-neutral price process of the underlying is assumed to be a martingale, i.e. the main driving process in Eq.(5.1) doesn't have a drift. However, the drift may not be negligible. If the drift is deterministic, e.g when the interest rate and dividends are deterministic, and the drift is a deterministic function of them, the calibration problem can be reduced to the driftless case by discounting, but this assumption might be inconsistent with the market. Therefore, an expansion of the proposed model that allows for a non-zero and stochastic drift is very desirable. In particular, it would be interesting to expand the LVG model to a risk-neutral price process obtained by stochastic time change of a drifted diffusion. In this way, similar to local Variance Gamma model, [Madan *et al.* (1998)], we introduce both stochastic volatility and stochastic drift.

With this in mind, en expanded version of the LVG model was proposed in [Carr and Itkin (2018a)] by adding drift to the governing underlying process. It turned out that this relatively simple step (at the first glance) requires re-derivation and re-thinking of almost every step in the construction proposed in [Carr and Nadtochiy (2017)]. The authors show that still it is possible to find an ordinary differential equation (ODE) for the option price which plays a role of Dupire's equation for the standard local volatility model, and how calibration of multiple smiles (the whole local volatility surface) can be done in such a case.

Further, assuming the local variance to be a piecewise linear function of strike and piecewise constant function of time we solve this ODE in closed form in terms of Confluent hypergeometric functions. Calibration of the model to market smiles does not require solving any optimization problem. In contrast, it can be done term-by-term by solving a system of non-linear algebraic equations for each maturity, and thus is much faster.

Below in this Chapter we describe the approach of [Carr and Itkin (2018a)] providing all the details as well as the results of computational experiments. The Chapter is organized as follows. In Section 5.1 the Expanded Local Variance Gamma model is formulated. In Section 5.2 we derive a forward equation (which is an ordinary differential equation (ODE)) for Put option prices using a homogeneous Bochner subordination approach. Section 5.3 generalizes this approach by considering the local variance being piece-wise constant in time. In Section 5.4 a closed form solution of the derived ODE is given in terms of Confluent hypergeometric functions. The next Section discusses computation of a source term of this ODE which requires a no-arbitrage interpolation. Using the idea of [Itkin and Lipton (2018)]), we show how to construct non-linear interpolation which provides both no-arbitrage, and a nice tractable representation of the source term, so that all integrals in the source term can be computed in closed form. In Section 5.6 calibration of multiple smiles in our model is discussed in detail. To calibrate a single smile we derive a system of nonlinear algebraic equations for the model parameters, and explain how to obtain a smart guess for their initial values. In Section 5.7 asymptotic solutions of our ODE at extreme values of the model parameters are derived which improve computational accuracy and speed of the numerical solution. Section 5.8 presents the results of some numerical experiments where calibration of the model to the given market smiles is done term-by-term.

5.1 Process

Below where possible we follow the notation of [Carr and Nadtochiy (2017)].

Let W_t be a $\mathbb{Q}$ standard Brownian motion with time index $t \geq 0$. Consider a stochastic process D_t to be a time-homogeneous diffusion

$$dD_t = \mu D_t dt + \sigma(D_t) dW_t, \tag{5.1}$$

where the volatility function σ is local and time-homogeneous, and μ is deterministic.

A unique solution to Eq.(5.1) exists if $\sigma(D) : \mathbb{R} \to \mathbb{R}$ is Lipschitz continuous in D and satisfies growth conditions at infinity. According to Eq.(5.1) we have $D_t \in (-\infty, \infty)$ while $t \in [0, \infty)$. Since D is a time-homogeneous Markov process, its infinitesimal generator $\mathcal{A}$ is given by

$$\mathcal{A}\phi(D) \equiv \left[\mu \dot{D} \nabla_D + \frac{1}{2} \sigma^2(D) \nabla_D^2 \right] \phi(D) \tag{5.2}$$

for all twice differentiable functions ϕ. Here ∇_x is a first order differential operator on x. The semigroup of the D process is

$$\mathcal{T}_t^D \phi(D_t) = e^{t\mathcal{A}} \phi(D_t) = \mathbb{E}_\mathbb{Q}[\phi(D_t)|D_0 = D], \quad \forall t \geq 0. \tag{5.3}$$

In the spirit of Variance Gamma model, [Madan and Seneta (1990); Madan *et al.* (1998)] and similar to [Carr and Nadtochiy (2017)], introduce a new process D_{Γ_t} which is D_t subordinated by the unbiased Gamma clock Γ_t. The density of the unbiased Gamma clock Γ_t at time $t \geq 0$ is

$$\mathbb{Q}\{\Gamma_t \in d\nu\} = \frac{\nu^{m-1} e^{-\nu m/t}}{(t^*)^m \Gamma(m)} d\nu, \quad \nu > 0, \quad m \equiv t/t^*. \tag{5.4}$$

Here $t^* > 0$ is a free parameter of the process, $\Gamma(x)$ is the Gamma function. It is easy to check that

$$\mathbb{E}_\mathbb{Q}[\Gamma_t] = t. \tag{5.5}$$

Thus, on average the stochastic gamma clock Γ_t runs synchronously with the calendar time t.

As applied to the option pricing problem, we introduce a more complex construction. Namely, consider options written on the underlying process S_t. Without loss of generality and for the sake of clearness let us treat below S_t as the stock price process. Here, in contrast to [Carr and Nadtochiy (2017)], we don't ignore interest rates r and continuous dividends q assuming them to be deterministic (below for simplicity of presentation we treat them as constants, but this can be easily relaxed). Then, let us define S_t as

$$S_t = D_{\Gamma_{X(t)}}, \qquad X(t) = \frac{1 - e^{-(r-q)t}}{r - q}. \tag{5.6}$$

It is clear that in the limit $r \to 0$, $q \to 0$ we have $X(t) = t$, i.e., in this limit our construction coincides with that in [Carr and Nadtochiy (2017)] who considered a driftless diffusion and assumed $S_t = D_{\Gamma_t}$. Also based on Eq.(5.5)

$$\mathbb{E}_\mathbb{Q}[\Gamma_{X(t)}] = X(t). \tag{5.7}$$

Function $X(t)$ starts at zero, i.e., $X(0) = 0$, and is a continuous increasing function of time t. Indeed, if $r - q > 0$, then $X(t)$ is increasing in t on $t \in [0, \infty)$, and at $t \to \infty$ it tends to constant. The infinite time horizon is not practically important, but for any finite time t function $X(t)$ can be treated as an increasing function in t. If $r - q < 0$, function $X(t)$ is strictly increasing $\forall t \in [0, \infty)$. Thus, $X(t)$ has all properties of a good clock. Accordingly, $\Gamma_{X(t)}$ has all properties of a random time.

Under a risk-neutral measure $\mathbb{Q}$, the total gain process, including the underlying price appreciation and dividends, after discounting at the risk free rate should be a martingale, see, e.g., [Shreve (1992)]. This process obeys the following stochastic differential equation

$$d\left(e^{-rt}S_t e^{qt}\right) = e^{(q-r)t}\left[(q-r)S_t dt + dS_t\right]. \tag{5.8}$$

Taking an expectation of both parts we obtain

$$\mathbb{E}_{\mathbb{Q}}[d\left(e^{(q-r)t}S_t\right)] = e^{(q-r)t}\left\{(q-r)\mathbb{E}_{\mathbb{Q}}[S_t]dt + d\mathbb{E}_{\mathbb{Q}}[S_t]\right\}. \tag{5.9}$$

Observe, that from Eq.(5.6), Eq.(5.1)

$$\begin{aligned}
\mathbb{E}_{\mathbb{Q}}[dS_t] &= \mathbb{E}_{\mathbb{Q}}[dD_{\Gamma_{X(t)}}] = \mu\mathbb{E}_{\mathbb{Q}}[D_{\Gamma_{X(t)}}d\Gamma_{X(t)}] + \mathbb{E}_{\mathbb{Q}}[\sigma(D_{\Gamma_{X(t)}})dW_{\Gamma_{X(t)}}] \\
&= \mu\mathbb{E}_{\mathbb{Q}}[D_{\Gamma_{X(t)}}d\Gamma_{X(t)}], \tag{5.10}
\end{aligned}$$

because the process W_{Γ_t} is a local martingale, see [Revuz and Yor (1999)], chapter 6. Accordingly, the process $W_{\Gamma_{X(t)}}$ inherits this property from W_{Γ_t}, hence $\mathbb{E}_{\mathbb{Q}}[\sigma(D_{\Gamma_{X(t)}})dW_{\Gamma_{X(t)}}] = 0$.

Further assume that the Gamma process Γ_t is independent of W_t (and, accordingly, $\Gamma_{X(t)}$ is independent of $W_{\Gamma_{X(t)}}$). Then the expectation in the RHS of Eq.(5.10) can be computed, by first conditioning on $\Gamma_{X(t)}$, and then integrating over the distribution of $\Gamma_{X(t)}$ which can be obtained from Eq.(5.4) by replacing t with $X(t)$, i.e.

$$\begin{aligned}
\mathbb{E}_{\mathbb{Q}}[D_{\Gamma_{X(t)}}d\Gamma_{X(t)}|S_s] &= \int_0^\infty \mathbb{E}_{\mathbb{Q}}[D_{\Gamma_{X(t)}}d\Gamma_{X(t)}|\Gamma_{X(t)} = \nu]\frac{\nu^{m-1}e^{-\nu m/X(t)}}{(t^*)^m\Gamma(m)} \\
&= \int_0^\infty \mathbb{E}_{\mathbb{Q}}[D_\nu]\frac{\nu^{m-1}e^{-\nu m/X(t)}}{(t^*)^m\Gamma(m)}d\nu, \tag{5.11} \\
\nu > 0, \quad &m \equiv X(t)/t^*.
\end{aligned}$$

The find $\mathbb{E}_{\mathbb{Q}}[D_\nu]$ we take into account Eq.(5.1) to obtain

$$d\mathbb{E}_{\mathbb{Q}}[D_\nu] = \mathbb{E}_{\mathbb{Q}}[dD_\nu] = \mathbb{E}_{\mathbb{Q}}[\mu D_\nu d\nu + \sigma(D_\nu)D_\nu dW_\nu] = \mu\mathbb{E}_{\mathbb{Q}}[D_\nu]d\nu. \tag{5.12}$$

Solving this equation with respect to $y(\nu) = \mathbb{E}_{\mathbb{Q}}[D_\nu|D_s]$, we obtain $\mathbb{E}_{\mathbb{Q}}[D_\nu|D_s] = D_s e^{\mu(\nu-s)}$. Since we condition on time s, it means that $D_s = D_{\Gamma_{X(s)}} = S_s$, and thus $\mathbb{E}_{\mathbb{Q}}[D_\nu|D_s] = S_s e^{\mu(\nu-s)}$.

Further, we substitute this into Eq.(5.11), set the parameter of the Gamma distribution t^* to be $t^* = X(t)$ (so $m = 1$) and integrate to obtain

$$d\mathbb{E}_{\mathbb{Q}}[S_t|S_s] = \mathbb{E}_{\mathbb{Q}}[dS_t|S_s] = \mu\mathbb{E}_{\mathbb{Q}}[D_{\Gamma_{X(t)}}d\Gamma_{X(t)}] = S_s\frac{e^{-s\mu}\mu}{1 - \mu X(t)}. \tag{5.13}$$

Setting now $m = r - q$ and solving this equation we find

$$\mathbb{E}_{\mathbb{Q}}[S_t|S_s] = S_s(r - q)e^{(q-r)(s-t)}. \tag{5.14}$$

Substituting Eq.(5.14) and Eq.(5.13) into Eq.(5.9) yields $d\left(e^{-rt}S_t e^{qt}\right) = 0$. Thus, if we chose $\mu = r - q$, the right hands part of Eq.(5.8) vanishes, and our discounted stock process with allowance for non-zero interest rates and continuous dividends becomes a martingale. So the proposed construction can be used for option pricing.

This setting can be easily generalized for time-dependent interest rates $r(t)$ and continuous dividends $q(t)$. We leave it for the reader.

The next step is to consider connection between the original and time-changed processes. It is known from [Bochner (1949)] that the process G_{Γ_t} defined as

$$dG_t = \sigma^2(G)dW_t$$

is a time-homogeneous Markov process. As the deterministic process μt is also time-homogeneous, the whole process D_t defined in Eq.(5.1) is also a time-homogeneous Markov process. Accordingly, the semigroups T_t^S of S_t and T_t^D of $D_{\Gamma_{X(t)}}$ are connected by the Bochner integral

$$\mathcal{T}_t^S U(S) = \int_0^\infty \mathcal{T}_\nu^D U(S)\mathbb{Q}\{\Gamma_{X(t)} \in d\nu\}, \quad \forall t \geq 0, \qquad (5.15)$$

where $U(S)$ is a function in the domain of both $\mathcal{T}_t^D$ and $\mathcal{T}_t^S$. It can be derived by exploiting the time homogeneity of the D process, conditioning on the gamma time first, and taking into account the independence of Γ_t and W_t (or $\Gamma_{\Gamma_{X(t)}}$ and $W_{\Gamma_{X(t)}}$ in our case).

We set parameter t^* of the gamma clock to $t^* = X(t)$. Then Eq.(5.15) and Eq.(5.4) imply

$$\mathcal{T}_t^S U(S) = \int_0^\infty \mathcal{T}_\nu^D U(S)\frac{e^{-\nu/X(t)}}{X(t)}d\nu. \qquad (5.16)$$

In what follows for the sake of brevity we will call this model as an Expanded Local Variance Gamma model, or ELVG.

5.2 Forward equation for Put option prices

Following [Carr and Nadtochiy (2017)] we interpret the index t of the semigroup $\mathcal{T}_t^S$ as the maturity date T of a European claim with the valuation time $t = 0$. Also let the test function $U(S)$ be the payoff of this European claim, i.e.,

$$U(S_T) = e^{-rT}(K - S_T)^+. \qquad (5.17)$$

Then define

$$P(S_0, T, K) = \mathcal{T}_T^S U(S_0), \qquad (5.18)$$

as the European Put value with maturity T at time $t = 0$ in the ELVG model. Similarly

$$P^D(S_0, \nu, K) = \mathcal{T}_\nu^D U(S_0),\qquad (5.19)$$

would be the European Put value with maturity ν at time $t = 0$ in the model of Eq.(5.1).[1] Then the Bochner integral in Eq.(5.16) takes the form

$$P(S, T, K) = \int_0^\infty P^D(S, \nu, K)pe^{-p\nu}d\nu, \quad p \equiv 1/X(T). \qquad (5.20)$$

Thus, $P(S, X(T), K)$ is represented by a Laplace-Carson transform of $P^D(S, \nu, K)$ with p being a parameter of the transform. Note that

$$P(S, 0, K) = P^D(S, 0, K) = U(S). \qquad (5.21)$$

To proceed, we need an analog of the Dupire forward PDE for $P^D(S, \nu, K)$.

5.2.1   *Derivation of the Dupire forward PDE*

Despite this can be done in many different ways, below for the sake of compatibility we do it in the spirit of [Carr and Nadtochiy (2017)]. First, differentiating Eq.(5.19) by ν with allowance for Eq.(5.3) yields

$$\nabla_\nu P^D(S, \nu, K) = e^{-r\nu}e^{\nu\mathcal{A}}\left[\mathcal{A} - r\right]U(S) = e^{-r\nu}\mathbb{E}_{\mathbb{Q}}\left[\mathcal{A} - r\right]U(S). \qquad (5.22)$$

We take into account the definition of the generator $\mathcal{A}$ in Eq.(5.2), and also remind that at $t = 0$ we have $D_0 = S_0$. Then Eq.(5.22) transforms to

$$\nabla_\nu P^D(S, \nu, K) = -rP^D(S, \nu, K) + (r - q)S\nabla_S P^D(S, \nu, K) \qquad (5.23)$$
$$+ e^{-r\nu}\frac{1}{2}\mathbb{E}_{\mathbb{Q}}\left[\sigma^2(S)\nabla_S^2 U(S)\right].$$

However, we need to express the forward equation using a pair of independent variables (ν, K) while Eq.(5.22) is derived in terms of (ν, S). To do this, observe that

$$e^{-r\nu}\mathbb{E}_{\mathbb{Q}}\left[\sigma^2(S)\nabla_S^2 U(S)\right] = e^{-r\nu}\mathbb{E}_{\mathbb{Q}}\left[\sigma^2(S)\delta(K - S)\right] \qquad (5.24)$$
$$= e^{-r\nu}\mathbb{E}_{\mathbb{Q}}\left[\sigma^2(K)\delta(K - S)\right] = e^{-r\nu}\mathbb{E}_{\mathbb{Q}}\left[\sigma^2(K)\nabla_K^2 U(S)\right]$$
$$= \sigma^2(K)\nabla_K^2 P^D(S, \nu, K).$$

---

[1]Below for simplicity of notation we drop the subscript '0' in S_0.

where the sifting property of the Dirac delta function $\delta(S - K)$ has been used. Also

$$-rP^D(S, \nu, K) + (r - q)S\nabla_S P^D(S, \nu, K) \tag{5.25}$$

$$= e^{-r\nu}\mathbb{E}_{\mathbb{Q}}\left[-r(K - S)^+ + (r - q)S\frac{\partial(K - S)^+}{\partial S}\right]$$

$$= e^{-r\nu}\mathbb{E}_{\mathbb{Q}}\left[-r(K - S)^+ - (r - q)(K - S)\frac{\partial(K - S)^+}{\partial S}\right.$$

$$\left. + (r - q)K\frac{\partial(K - S)^+}{\partial S}\right]$$

$$= e^{-r\nu}\mathbb{E}_{\mathbb{Q}}\left[-r(K - S)^+ + (r - q)(K - S)^+ - (r - q)K\frac{\partial(K - S)^+}{\partial K}\right]$$

$$= -qP^D(S, \nu, K) - (r - q)K\nabla_K P^D(S, \nu, K).$$

Therefore, using Eq.(5.24) and Eq.(5.25), Eq.(5.22) could be transformed to

$$\nabla_\nu P^D(S, \nu, K) = -qP^D(S, \nu, K) - (r - q)K\nabla_K P^D(S, \nu, K)$$

$$+ \frac{1}{2}\sigma^2(K)K^2\nabla_K^2 P^D(S, \nu, K) \equiv \mathcal{A}^K P^D(S, \nu, K), \tag{5.26}$$

$$\mathcal{A}^K = -q - (r - q)K\nabla_K + \frac{1}{2}\sigma^2(K)K^2\nabla_K^2.$$

This equation looks exactly like the Dupire equation with non-zero interest rates and continuous dividends, see, e.g., [Ekström and Tysk (2012)] and references therein. Note, that $\mathcal{A}^K$ is also a time-homogeneous generator.

5.2.2   *Forward partial divided-difference equation*

Our final step is to apply the linear differential operator $\mathcal{A}^K$ defined in Eq.(5.26) to both parts of Eq.(5.20). Using time-homogeneity of D_t and, again, the Dupire equation Eq.(5.26), we obtain

$$-qP(S, T, K) - (r - q)K\nabla_K P(S, T, K) + \frac{1}{2}\sigma^2(K)\nabla_K^2 P(S, T, K) \tag{5.27}$$

$$= \int_0^\infty pe^{-p\nu}\left[-qP^D(S, \nu, K) - (r - q)K\nabla_K P^D(S, \nu, K)\right.$$

$$\left. + \frac{1}{2}\sigma^2(K)\nabla_K^2 P^D(S, \nu, K)\right]d\nu = \int_0^\infty pe^{-p\nu}\nabla_\nu P^D(S, \nu, K)d\nu$$

$$= -pP^D(S, 0, K) + p\int_0^\infty P^D(S, \nu, K)pe^{-p\nu}d\nu$$

$$= p\left[P(S, T, K) - P^D(S, 0, K)\right] = p\left[P(S, T, K) - P(S, 0, K)\right],$$

where in the last line Eq.(5.21) was taken into account.

Thus, finally $P(S, T, K)$ solves the following problem

$$-qP(S,T,K) - (r-q)K\nabla_K P(S,T,K) + \frac{1}{2}\sigma^2(K)\nabla_K^2 P(S,T,K) \quad (5.28)$$

$$= \frac{P(S,T,K) - P(S,0,K)}{X(T)}, \qquad P(S,0,K) = (K-S)^+.$$

At $r = q = 0$ this equation translates to the corresponding equation in [Carr and Nadtochiy (2017)]. In contrast to the Dupire equation which belongs to the class of PDE, Eq.(5.28) is an ODE, or, more precisely, a partial divided-difference equation (PDDE), since the derivative in time in the right hands part is now replaced by a divided difference. In the form of an ODE it reads

$$\left[\frac{1}{2}\sigma^2(K)\nabla_K^2 - (r-q)K\nabla_K - \left(q + \frac{1}{X(T)}\right)\right]P(S,T,K) = -\frac{P(S,0,K)}{X(T)}.$$
$$(5.29)$$

This equation could be solved analytically for some particular forms of the local volatility function $\sigma(K)$ which are considered later in this Chapter. Also in the same way a similar equation could be derived for the Call option price $C_0(S, T, K)$ which reads

$$\left[\frac{1}{2}\sigma^2(K)\nabla_K^2 - (r-q)K\nabla_K - \left(q + \frac{1}{X(T)}\right)\right]C_0(S,T,K) = -\frac{C_0(S,0,K)}{X(T)},$$
$$C_0(S,0,K) = (S-K)^+. \qquad (5.30)$$

Solving Eq.(5.29) or Eq.(5.30) provides the way to determine $\sigma(K)$ given market quotes of Call and Put options with maturity T. However, this allows calibration of just a single term. Calibration of the whole local volatility surface, in principle, could be done term-by-term (because of the time-homogeneity assumption) if Eq.(5.29), Eq.(5.30) could be generalized to this case. We consider this in the following Section.

5.3   Local variance piece-wise constant in time

To address calibration of multiple smiles, we need to relax some assumptions about time-homogeneity of the process D_t defined in Eq.(5.1). This includes several steps which are described below in more detail.

5.3.1   *Local variance*

Here we assume that the local variance $\sigma(D_t)$ is no more time-homogeneous, but a piece-wise constant function of time $\sigma(D_t, t)$.

Let $T_1, T_2, \ldots, T_M$ be the time points at which the variance rate $\sigma^2(D_t)$ jumps deterministically. In other words, at the interval $t \in [T_0, T_1)$, the variance rate is $\sigma_0^2(D_t)$, at $t \in [T_1, T_2)$ it is $\sigma_1^2(D_t)$, etc. This can be also represented as

$$\sigma^2(D_t, t) = \sum_{i=0}^{M} \sigma_i^2(D_t) w_i(t), \tag{5.31}$$

$$w_i(t) \equiv \mathbf{1}_{t-T_i} - \mathbf{1}_{t-T_{i+1}}, \quad i = 0, \ldots, M, \quad T_0 = 0, \quad T_{M+1} = \infty,$$

$$\mathbf{1}_x = \begin{cases} 1, & x \geq 0 \\ 0, & x < 0. \end{cases}$$

Note, that

$$\sum_{i=0}^{M} w_i(t) = \mathbf{1}_t - \mathbf{1}_{t-\infty} = 1, \quad \forall t \geq 0.$$

Therefore, in case when all $\sigma_i^2(D_t)$ are equal, i.e., independent on index i, Eq.(5.31) reduces to the case considered in the previous Sections.

It is important to notice that our construction implies that the volatility $\sigma(D_t)$ jumps as a function of time at the calendar times $T_0, T_1, \ldots, T_M$, and not at the business times ν determined by the gamma clock. Otherwise, the volatility function would change at random (business) times which means it is stochastic. But this definitely lies out of scope of our model. Therefore, we need to change Eq.(5.31) to

$$\sigma^2(D_{\Gamma_{X(t)}}, \Gamma_{X(t)}) = \sum_{i=0}^{M} \sigma_i^2(D_t) \bar{w}_i(\mathbb{E}_{\mathbb{Q}}(\Gamma_{X(t)})), \tag{5.32}$$

$$\bar{w}_i(t) = \mathbf{1}_{X^{-1}(t)-T_i} - \mathbf{1}_{X^{-1}(t)-T_{i+1}}, \quad i = 0, \ldots, M,$$

$$X^{-1}(t) = \frac{1}{q-r} \log\left[1 - (r-q)t\right].$$

Hence, when using Eq.(5.6) we have

$$\sigma^2(D_t, t)\Big|_{t=\Gamma_{X(t)}} = \sum_{i=0}^{M} \sigma_i^2(D_t) \bar{w}_i(X(t)) = \sum_{i=0}^{M} \sigma_i^2(D_t) w_i(t). \tag{5.33}$$

Accordingly, if the calendar time t belongs to the interval $T_0 \leq t < T_1$, the infinitesimal generator $\mathcal{A}$ of the semigroup $\mathcal{T}_\nu^D$ is a function of $\sigma(D_t)$ (and not on $\sigma(D_\nu)$). As at $T_0 \leq t < T_1$ we assume $\sigma(D) = \sigma_0(D)$, i.e., is constant in time, it doesn't depend of ν. Thus, $\mathcal{A}$ (which for this interval of time we will denote as $\mathcal{A}_0$) is still time-homogeneous.

Similarly, one can see, that for $T_1 \leq t < T_2$ the infinitesimal generator $\mathcal{A}_1$ of the semigroup $\mathcal{T}_\nu^D$ is also time-homogeneous and depends on $\sigma_1(D)$, etc.

5.3.2 Bochner subordination

We start with re-definition of Eq.(5.18), Eq.(5.19). We now define the European Put value with maturity T at the evaluation time $t = X(T_1)$ in the ELVG model

$$P(S_0, T_1 + T, K) = \mathcal{T}_T^S[e^{-rT}P(S_0, T_1, K)]. \tag{5.34}$$

And, clearly we are interesting in the value of T to be $T = T_2 - T_1$.

Similarly, we define the European Put value with maturity ν at the evaluation time $t = T_1$ in the model given by Eq.(5.1) as

$$P^D(S_0, T_1 + \nu, K) = \mathcal{T}_\nu^D[e^{-r\nu}P(S_0, T_1, K)]. \tag{5.35}$$

By these definitions

$$P(S_0, T_1 + T, K)\Big|_{T=0} = P^D(S_0, T_1 + \nu, K)\Big|_{\nu=0} = P(S_0, T_1^\bullet, K)].$$

In contrast to Eq.(5.20), in case of multiple smiles at $t > T_1$ we need to change the definition of t in Eq.(5.16) from $t \mapsto X(t)$ to

$$t \mapsto X(T_1 + t) - X(T_1) \equiv \Delta x(T_1, t). \tag{5.36}$$

This definition implies two observations.

First, function $\Delta x(T_1, t)$ starts at zero at $t = 0$ and is an increasing function of time. Also, in case $r = q = 0$ we have $\Delta x(T_1, t) = t$. Therefore, $\Delta x(T_1, t)$ can be used as a good clock. Accordingly, similar to Eq.(5.5) we have

$$\mathbb{E}_{\mathbb{Q}}[\Gamma_{\Delta x(T_1,t)}] = \Delta x(T_1, t). \tag{5.37}$$

Second, a proof that in our model the discounted stock price is a martingale given in Section 5.1 could be repeated for times $t: T_1 < t \leq T_2$. When doing so, at $t > T_1$ we reset the definition of S_t to

$$S_{T_1+t} = D_{\Gamma_{\Delta x(T_1,t)}}, \quad t \geq 0.$$

Then instead of Eq.(5.10) we now have

$$\mathbb{E}_{\mathbb{Q}}[dS_{T_1+t}] = \mathbb{E}_{\mathbb{Q}}[dD_{\Gamma_{\Delta x(T_1,t)}}] \tag{5.38}$$

$$= \mu\mathbb{E}_{\mathbb{Q}}[D_{\Gamma_{\Delta x(T_1,t)}}d\Gamma_{\Delta x(T_1,t)}] + \mathbb{E}_{\mathbb{Q}}[\sigma(D_{\Gamma_{\Delta x(T_1,t)}})dW_{\Gamma_{\Delta x(T_1,t)}}]$$

$$= \mu\mathbb{E}_{\mathbb{Q}}[D_{\Gamma_{\Delta x(T_1,t)}}]d\Delta x(T_1, t) = \mu\mathbb{E}_{\mathbb{Q}}[D_{\Gamma_{\Delta x(T_1,t)}}]dX(T_1 + t).$$

On the other hand,

$$\mathbb{E}_{\mathbb{Q}}[d\left(e^{(q-r)(T_1+t)}S_{T_1+t}\right)] = e^{(q-r)(T_1+t)}\{(q-r)\mathbb{E}_{\mathbb{Q}}[S_{T_1+t}]dt + d\mathbb{E}_{\mathbb{Q}}[S_{T_1+t}]\}$$

$$= e^{(q-r)t}[\mu + (q-r)S_{T_1}e^{-(r-q)T_1}]dt. \tag{5.39}$$

One can check, that with $\mu = r - q$ the RHS of Eq.(5.39) vanishes, therefore this construction can be used for option pricing.

The definition in Eq.(5.36) implies that parameter t of the Gamma random clock is reset at the point T_1, i.e., at $0 \leq t \leq T_1$ it is $t \mapsto X(t) = X(t) - X(0)$, while at $T_1 < t \leq T_2$ it is $t \mapsto X(T_1 + t) - X(T_1)$. Using the definition of $w_i(t)$ in Eq.(5.31), this could be written as

$$t \mapsto \sum_{i=0}^{M} w_i(T_i + t)[X(T_i + t) - X(T_i)]. \tag{5.40}$$

Resetting t was also first proposed in [Carr and Nadtochiy (2017)] but in a different form.

Then, the Bochner integral in Eq.(5.16) transforms to

$$\mathcal{T}_T^S P(S, \overset{\bullet}{T}_1, K) = \int_0^\infty \mathcal{T}_\nu^D P(S, T_1 + \nu, K) \frac{\nu^{m-1} e^{-\nu m / \Delta X(T_1, T)}}{(t^*)^m \Gamma(m)} d\nu. \tag{5.41}$$

Since for a tractability reason we still want to have $m \equiv \Delta X(T_1, T)/t^* = 1$. we need to redefine t^* in accordance with Eq.(5.40). Based on that, the Bochner integral in Eq.(5.20) now finally reads

$$P(S, T_1 + T, K) = \int_0^\infty P^D(S, T_1 + \nu, K) p e^{-p\nu} d\nu, \quad p \equiv 1/\Delta X(T_1, T). \tag{5.42}$$

5.3.3 *Forward partial divided-difference equation for the second term*

Now we need to derive a Forward partial divided-difference equation for the second term T_2 similar to how this is done in Section 5.2.2. Obviously, the Put price $P^D(S_0, T_1 + \nu, K)$ solves the same Dupire equation Eq.(5.26). Therefore, proceeding in the same way as in Section 5.2.2, we apply linear differential operator $\mathcal{L}$ defined in Eq.(5.26) to both parts of Eq.(5.42). Using time-homogeneity of D_t at the interval $[T_1, T_2]$ and again the Dupire equation Eq.(5.26), we obtain

$$-qP(S, T_1 + T, K) - (r - q)K\nabla_K P(S, T_1 + T, K)$$

$$+ \frac{1}{2}\sigma^2(K)\nabla_K^2 P(S, T_1 + T, K) \tag{5.43}$$

$$= \int_0^\infty p e^{-p\nu} \Big[-qP^D(S, T_1 + \nu, K) - (r - q)K\nabla_K P^D(S, T_1 + \nu, K)$$

$$+ \frac{1}{2}\sigma^2(K)\nabla_K^2 P^D(S, T_1 + \nu, K) \Big] d\nu = \int_0^\infty p e^{-p\nu} \nabla_\nu P^D(S, T_1 + \nu, K) d\nu$$

$$= -pP^D(S, T_1, K) + p \int_0^\infty P^D(S, T_1 + \nu, K)pe^{-p\nu}d\nu$$
$$= p\left[P(S, T_1 + T, K) - P^D(S, T_1, K)\right]$$
$$= p\left[P(S, T_1 + T, K) - P(S, T_1, K)\right].$$

Finally, taking $T = T_2 - T_1$ we obtain an ODE for the Put price $P(S, T_2, K)$.

$$\left[\frac{1}{2}\sigma^2(K)\nabla_K^2 - (r - q)K\nabla_K - \left(q + \frac{1}{X(T_2) - X(T_1)}\right)\right]P(S, T_2, K)$$
$$= -\frac{P(S, T_1, K)}{X(T_2) - X(T_1)}. \tag{5.44}$$

Here the local variance function $\sigma^2(K) = \sigma_1^2(K)$ as it corresponds to the interval $(T_1, T_2]$ where the above ODE is solved.

We continue in the same way to derive an ODE for the Put price $P(S, T_i, K)$, $i = 1, \ldots, M$, which finally reads

$$\left[\frac{1}{2}\sigma^2(K)\nabla_K^2 - (r - q)K\nabla_K - \left(q + \frac{1}{X(T_i) - X(T_{i-1})}\right)\right]P(S, T_i, K)$$
$$= -\frac{P(S, T_{i-1}, K)}{X(T_i) - X(T_{i-1})}. \tag{5.45}$$

This is a recurrent equation that can be solved for all $i = 1, \ldots, M$ sequentially starting with $i = 1$ subject to some boundary conditions. The natural boundary conditions for the Put option price are, [Hull (1997)]

$$\begin{aligned} P(S, T_i, K) &= 0, & K &\to 0, \\ P(S, T_i, K) &= \mathcal{D}_i K - \mathcal{Q}_i S \approx \mathcal{D}_i K, & K &\to \infty, \end{aligned} \tag{5.46}$$

where $\mathcal{D}_i = e^{-rT_i}$ is the discount factor, and $\mathcal{Q}_i = e^{-qT_i}$.

A similar equation can be obtained for the Call option prices, which reads

$$\left[\frac{1}{2}\sigma^2(K)\nabla_K^2 - (r - q)K\nabla_K - \left(q + \frac{1}{X(T_i) - X(T_{i-1})}\right)\right]C(S, T_i, K)$$
$$= -\frac{C(S, T_{i-1}, K)}{X(T_i) - X(T_{i-1})}, \tag{5.47}$$

subject to the boundary conditions

$$\begin{aligned} C(S, T_i, K) &= \mathcal{Q}_i S, & K &\to 0, \\ C(S, T_i, K) &= 0, & K &\to \infty. \end{aligned} \tag{5.48}$$

A more careful analysis shows that the above boundary conditions are not rigorous, while are a good approximation of a the real boundary conditions. This analysis is presented in the next Chapter in Section 6.2.4. Since our equations are piecewise constant in time, i.e. discrete, the boundary conditions turns out to be also discrete. But they converge to the standard boundary conditions in the continuous case (Dupire). These conditions are constructed using some analog of discrete compounding which is natural for the LVG model, again see Section 6.2.4.

5.4   Solution of the ODE Eq.(5.45)

Below we use the approach similar to [Itkin and Lipton (2018)] by assuming the local variance to be a piecewise linear continuous function of strike. In contrast to [Itkin and Lipton (2018)], instead of a standard local volatility model we use the ELVG model. As the result, instead of a partial differential (Dupire) equation, we face a problem of solving the ODE in Eq.(5.45).

It is worth noticing, however, that for any piecewise model of the local variance/volatility, at edge intervals where strikes are close either to 0 or to infinity one has to switch to the local variance linear in the log-strike because of Roger Lee's moment formula, [Lee (2004)]. Thus, the whole local variance/volatility model becomes a combination of the original model at the internal intervals and local variance linear in log-strike at the edge intervals. In this Chapter we neglect this to illustrate our construction in more transparent way. However, we will give the full consideration of this approach in Chapter 6.

We proceed by first, doing a change of the dependent variable from $P(S, T_j, K)$ to

$$V(S, T_j, K) = P(S, T_j, K) - \mathcal{D}_j K,$$

which is known as a *covered Put*. The advantage of the covered Put is that according to Eq.(5.46) its price obeys homogeneous boundary conditions.

Using this definition we now re-write Eq.(5.45) in a more convenient form (while with some loose of notation)

$$- v(x)V_{x,x}(x) + b_1 x V_x(x) + b_{0,j}V(x) = c_j, \tag{5.49}$$

$$b_1 = (r - q)p_j, \quad b_{0,j} = qp_j + 1, \quad c_j = V(T_{j-1}, x) + \beta x,$$

$$p_j = X(T_j) - X(T_{j-1}) > 0, \quad x = \frac{K}{S}, \quad V(x) = V(S, T_j, x),$$

$$v(x) = p_j \frac{\sigma^2(x)}{2S^2}, \beta = -S[\mathcal{D}_j(1 + p_j r) - \mathcal{D}_{j-1}].$$

In Eq.(5.49) x is the inverse moneyness. In what follows we also assume that $r > q > 0$, but this assumption could be easily relaxed.

Further, suppose that for each maturity T_j, $j \in [1, M]$ the market quotes are provided at a set of strikes K_i, $i = 1, \ldots, n_j$ where the strikes are assumed to be sorted in the increasing order. Then the corresponding continuous piecewise linear local variance function $v_j(x)$ on the interval $[x_i, x_{i+1}]$ reads

$$v_{j,i}(x) = v_{j,i}^0 + v_{j,i}^1 x, \qquad (5.50)$$

where we use the super-index 0 to denote a level v^0, and the super-index 1 to denote a slope v^1. Subindex $i = 0$ in $v_{j,0}^0, v_{j,0}^1$ corresponds to the interval $(0, x_1]$. Since $v_j(x)$ is continuous, we have

$$v_{j,i}^0 + v_{j,i}^1 x_{i+1} = v_{j,i+1}^0 + v_{j,i+1}^1 x_{i+1}, \quad i = 0, \ldots, n_j - 1. \qquad (5.51)$$

The first derivative of $v_j(x)$ experiences a jump at points x_i, $i \in \mathbb{Z} \cap [1, n_j]$. We also assume that $v(x, T)$ is a piecewise constant function of time, i.e., $v_{j,i}^0, v_{j,i}^1$ don't depend on T on the intervals $[T_j, T_{j+1})$, $j \in [0, M - 1]$, and jump to new values at the points T_j, $j \in \mathbb{Z} \cap [1, M]$.

With the above assumptions in mind, Eq.(5.49) can be solved by induction. One starts with $T_0 = 0$, and on each time interval $[T_{j-1}, T_j]$, $j \in \mathbb{Z} \cap [1, M]$ solves the problem Eq.(5.49) for $V(x) \mapsto P(S, T_j, x) - d_j S x$.

Since $v(x)$ is a piecewise linear function, the solution of Eq.(5.49) can also be constructed separately for each interval $[x_{i-1}, x_i]$. By taking into account the explicit representation of $v(x)$ in Eq.(5.50), from Eq.(5.49) for the i-th spatial interval we obtain

$$-(b_2 + a_2 x)V_{xx}(x) + b_1 x V_x(x) + b_0 V(x) = c \qquad (5.52)$$

$$b_2 = v_{j,i}^0, \ a_2 = v_{j,i}^1.$$

We proceed by introducing a new independent variable $z = (b_2 + a_2 x)b_1/a_2^2$, $z \in \mathbb{R}^+$, so that Eq.(5.52) transforms to

$$-z V_{zz}(z) + (z - q_2)V_z(z) + q_1 V(z) = \chi \qquad (5.53)$$

$$q_1 = b_0/b_1, \ q_2 = b_2 b_1/a_2^2, \ \chi = c/b_1.$$

The Eq.(5.53) is an *inhomogeneous* Laplace equation, [Polyanin and Zaitsev (2003)], page 155. It is well known that if $y_1 = y_1(z)$, $y_2 = y_2(z)$ are two fundamental solutions of the corresponding *homogeneous* equation, then the general solution of Eq.(5.53) can be represented as

$$V(z) = C_1 y_1(z) + C_2 y_2(z) + \frac{1}{b_1} I_{12}(z) \qquad (5.54)$$

$$I_{12}(z) = -y_2(z) \int \frac{y_1(z)f(z)}{W z} dz + y_1(z) \int \frac{y_2(z)f(z)}{W z} dz \equiv I_1 + I_2,$$

$$f(z) = V(T_{j-1}, z) - k_1 - k_2 z, \quad k_1 = \beta \frac{b_2}{a_2}, \quad k_2 = -\beta \frac{a_2}{b_1},$$

where $W = y_1(y_2)_z - y_2(y_1)_z$ is the so-called Wronskian, and β is defined in Eq.(5.49). Then the problem is to determine suitable fundamental solutions of the homogeneous Laplace equations. Based on [Polyanin and Zaitsev (2003)], if $a_2 \neq 0$, they read

$$y_i(z) = \mathcal{V}_i(q_1, q_2, z), \quad i = 1, 2 \tag{5.55}$$

Here $\mathcal{V}_i(a, b, z)$ is an arbitrary solution of the degenerate hypergeometric equation, i.e., Kummer's function, [Abramowitz and Stegun (1964)]. Two types of Kummer's functions are known, namely $M(a, b, z)$ and $U(a, b, z)$, which are Kummer's functions of the first and second kind.

It is known, that there exist several pairs of such independent solutions. Therefore, for every spatial interval in z among all possible fundamental pairs we have to determine just one which is numerically satisfactory at this interval (see [Olver (1997)] for the detailed definition of satisfactory solutions and the corresponding discussion). Since our boundary conditions are set at zero and positive infinity, we need a numerically satisfactory solution for the positive half of the real line.

Similar to [Itkin and Lipton (2018)], in the vicinity of the origin we choose the numerically satisfactory pair as, [Olver (1997)]

$$y_1(\chi) = M(q_1, q_2, z) = e^z M(q_2 - q_1, q_2, -z), \tag{5.56}$$
$$y_2(\chi) = z^{1-q_2} M(q_1 - q_2 + 1, 2 - q_2, z) = z^{1-q_2} e^z M(1 - q_1, 2 - q_2, -z),$$
$$W = \sin(\pi q_2) z^{-q_2} e^z / \pi.$$

However, in the vicinity of infinity the numerically satisfactory pair is, [Olver (1997)]

$$y_1(\chi) = U(q_1, q_2, z) = z^{1-q_2} U(q_1 - q_2 + 1, 2 - q_2, z), \tag{5.57}$$
$$y_2(\chi) = e^z U(q_2 - q_1, q_2, -z) = e^z z^{1-q_2} U(1 - q_1, 2 - q_2, -z),$$
$$W = (-1)^{q_1 - q_2} e^z z^{-q_2}.$$

As two solutions $J_1(q_1, q_2, z), J_2(q_1, q_2, z)$ are independent, Eq.(5.54) is a general solution of Eq.(5.53). Two constants C_1, C_2 should be determined based on the boundary conditions for the function $V(z)$.

The boundary conditions for the ODE Eq.(5.52) in a strike K space (or in x space) should be set at zero and infinity. Based on the usual shape of the local variance curve and its positivity, for $x \to 0$, we expect that $v_{j,i}^1 < 0$. Similarly, for $x \to \infty$ we expect that $v_{j,i}^1 > 0$. In between these two limits the local variance curve for a given maturity T_j is assumed to be continuous, but the slope of the curve could be both positive and negative, see, e.g., [Itkin (2015)] and references therein. Also, by definition $z = v_{j,i}$, and

Dom$(z) = \mathbb{R}^+$. Thus, at high strikes $a_2 = v^1_{j,i} > 0$. Therefore, the boundary conditions for Eq.(5.53) should be set at $z = b_2$ (which corresponds to the boundary $K = 0$) and at $z \to \infty$. These are the boundary conditions given in Eq.(5.46).

5.5 Computation of the source term

Computation of the source term pI_{12} in Eq.(5.54) could be achieved in several ways. The most straightforward one is to use numerical integration as the Put price $P(x, T_{i-1})$ as a function of x is already known when we solve Eq.(5.49) for $T = T_i$. However, as this is discussed in [Itkin and Lipton (2018)], and also in more detail in Chapter 2, function $P(x, T_{i-1})$ is known only for a discrete set of points in x. Therefore, some kind of interpolation is necessary to find its values at the other points.

5.5.1 *Computing the integrals in Eq.(5.54) far from $z = 0$*

Using the interpolation scheme described in Section 2.3, consider the first integral in Eq.(5.54). To remind, we compute it at some interval $z \in [z_i, z_{i+1}]$, $i \in \mathbb{Z} \cap [1, n_j]$. Picking together the solutions in Eq.(5.56) with the interpolation scheme for $P(z, T_{j-1})$ and Wronskians in Eq.(5.56), and substituting them into the first integral in Eq.(5.54) we obtain

$$\int \frac{y_2(z)f(z, T_{j-1})}{Wz}dz = A\Big[- B_0 + B_1 M(-2 - q_1, -1 - q_2, -z) \tag{5.58}$$
$$+ B_2 M(-1 - q_1, -q_2, -z) + B_3 M(-q_1, 1 - q_2, -z)\Big],$$

$$A = \frac{1}{b_1^2 q_1}\pi(1 - q_2)\csc(\pi q_2),$$

$$B_0 = \frac{1}{a_2^2(q_1 + 1)(q_1 + 2)}\Big[a_2 b_1(q_1 + 2)\left(a_2^2 \beta q_2 - b_1(q_1 + 1)(\beta b_2 - a_2 \gamma_1)\right)$$
$$+ \gamma_2 \left(2a_2^4 q_2(q_2 + 1) - 2a_2^2 b_1 b_2(q_1 + 2)q_2 + b_1^2 b_2^2(q_1 + 1)(q_1 + 2)\right)\Big],$$

$$B_1 = 2a_2^2 \gamma_2 \frac{q_2(q_2 + 1)}{(1 + q_1)(2 + q_1)},$$

$$B_2 = (a_2 b_1 \beta - 2b_1 b_2 \gamma_2 + 2a_2^2 \gamma_2 z)\frac{q_2}{1 + q_1},$$

$$B_3 = \frac{1}{a_2^2}\Big[a_2 b_1\left(a_2^2 \beta z + a_2 b_1 \gamma_1 - \beta b_1 b_2\right) + \gamma_2 \left(b_1 b_2 - a_2^2 z\right)^2\Big].$$

Similarly

$$\int \frac{y_1(z)f(z,T_{j-1})}{Wz}dz = \bar{A}\Big[\bar{B}_1 M(q_2 - q_1, 1 + q_2, -z)$$

$$+ \bar{B}_2 \; _2F_2\,(q_2 - q_1, 1 + q_2; q_2, 2 + q_2; -z)$$

$$+ \bar{B}_3 \; _2F_2\,(q_2 - q_1, 2 + q_2; q_2, 3 + q_2; -z)\Big], \qquad (5.59)$$

$$A = \pi z^{q_2}\csc(\pi q_2)\Gamma(q_2),$$

$$B_1 = \frac{a_2^2\gamma_1 - a_2\beta b_2 + b_2^2\gamma_2}{a_2^2\Gamma(1 + q_2)},$$

$$B_2 = \frac{\Gamma(q_2 + 1)}{\Gamma(q_2)\Gamma(2 + q_2)b_1}(a_2\beta - 2b_2\gamma_2)z,$$

$$B_3 = \frac{\Gamma(q_2 + 1)}{\Gamma(q_2)\Gamma(3 + q_2)b_1^2}a_2^2(1 + q_2)\gamma_2 z^2,$$

where $_pF_q\,(a_1, ..., a_p; b_1, ..., b_q; z)$ is the generalized hypergeometric function, [Olver (1997)].

5.5.2   *Computing the integrals in Eq.(5.54) far from*
$z = \pm\infty$

Here we proceed in the same way as in the previous section. Again, we pick together the solutions in Eq.(5.57) with the interpolation scheme for $P(z, T_{j-1})$ and Wronskians in Eq.(5.57), and substitute them into the first integral in Eq.(5.54) we obtain

$$\int \frac{y_2(z)f(z,T_{j-1})}{Wz}dz = (-1)^{q_2 - q_1}[C_0 J_0 + C_1 J_1 + C_2 J_2], \qquad (5.60)$$

$$J_i = \int z^i U(1 - q_1, 2 - q_2, -z)dz,$$

$$C_0 = \frac{b_2^2\gamma_2}{a_2^2} - \frac{\beta b_2}{a_2} + \gamma_1, \quad C_1 = \frac{a_2\beta - 2b_2\gamma_2}{b_1}, \quad C_2 = \frac{a_2^2\gamma_2}{b_1^2}.$$

It is known, [Abramowitz and Stegun (1964)], that

$$J_0 = -\frac{1}{q_1}U(-q_1, 1 - q_2, -z).$$

Then, J_1, J_2 can be found using integration by parts to yield

$$J_1 = zJ_0 + \frac{1}{q_1(1+q_1)}U(-1-q_1, -q_2, -z),$$

$$J_2 = zJ_1 + \frac{1}{q_1(2+3q_1+q_1^2)}U(-2-q_1, -1-q_2, -z)$$

$$- z\frac{1}{q_1(1+q_1)}U(-1-q_1, -q_2, -z).$$

Similarly

$$\int \frac{y_1(z)f(z, T_{j-1})}{Wz}dz = (-1)^{q_2-q_1}[C_0\mathcal{J}_0 + C_1\mathcal{J}_1 + C_2\mathcal{J}_2], \qquad (5.61)$$

$$\mathcal{J}_i = \int z^i e^{-z}U(1+q_1-q_2, 2-q_2, z)dz.$$

The integrals $\mathcal{J}_i$ have been considered in [Itkin and Lipton (2018)] using the approach of [Ng and Geller (1970)]. Borrowing from there the result

$$\mathcal{J}_0 = \int e^{-z}U(1+q_1-q_2, 2-q_2, z)dz = -e^{-z}U(q_1-q_2, 1-q_2, z),$$

and using integration by parts, we obtain

$$\mathcal{J}_1 = z\mathcal{J}_0 + e^{-z}U(q_1-q_2-1, -1-q_2, z),$$

$$\mathcal{J}_2 = z\mathcal{J}_1 - \int \mathcal{J}_1 dz = (z-1)\mathcal{J}_1 - \int e^{-z}U(q_1-q_2-1, -1-q_2, z)dz$$

$$= (z-1)\mathcal{J}_1 + e^{-z}U(q_1-q_2-2, -2-q_2, z).$$

5.5.3 *Some additional notes*

Based on the no-arbitrage interpolation and some analytics proposed in this Section, we managed to find the solution Eq.(5.54) of the forward equation Eq.(5.49) in closed form . This solution by construction is arbitrage free at any interval where the local variance function defined in Eq.(5.50) is linear. In other words we proved, that if we consider, say 3 strikes $0 < K_1 < K_2 < K_3 < \infty$ such that, e.g., $x_1 = K_1/S \in [x_i, x_{i+1}]$, $x_2 = K_2/S \in [x_i, x_{i+1}]$, $x_3 = K_3/S \in [x_i, x_{i+1}]$, then the solution at these 3 points obeys no-arbitrage conditions.

5.6 Calibration of smile for a given term T_i

Calibration problem for the local volatility model can be formulated as follows: given market quotes of Call and/or Put options corresponding to a set of N strikes $\{K\} := K_j$, $j \in [1, N]$ and same maturity T_i, find the local variance function $\sigma(K)$ such that these quotes solve equations in Eq.(5.45), Eq.(5.47).

As mentioned in [Itkin and Lipton (2018)], there are two main approaches to solving this problem. The first approach attempts to construct a continuous implied volatility (IV) surface matching the market quotes by using either some parametric or non-parametric regression, and then generates the corresponding LV surface via the well-known relationship between the local and implied variance s also known as the Dupire formula, see, e.g., [Itkin (2015)] and references therein. To be practically useful, this construction should guarantee no arbitrage for all strikes and maturities, which is a serious challenge for any model based on interpolation. If the no-arbitrage condition is satisfied, then the LV surface can be calculated using the Dupire formula. The second approach relies on the direct solution of the corresponding forward equation (which is the Dupire equation in the Black-Scholes world, or Eq.(5.47), Eq.(5.45) in our model) using either analytical or numerical methods. The advantage of this approach is that it guarantees no-arbitrage. However, the problem of the direct solution can be ill-posed, [Coleman *et al.* (2001)], and is rather computationally intensive.

In this Section we show that the second approach could be significantly simplified when using the ELVG model, so calibration of the smile could be done very fast and accurate.

Further, for the sake of certainty, suppose that all known market quotes are Puts, despite this can be easily relaxed. Also, suppose that the shape of a local variance is given by some function $\sigma_j(K) = f_j(K, p_1, \ldots, p_L)$, where $p_1, \ldots, p_L$ is a set of the model parameters to be determined. For instance, in [Lipton and Sepp (2011a); Carr and Nadtochiy (2017)] the local variance is assumed to be a piecewise constant function of strike, while in [Itkin and Lipton (2018)] this is a piecewise linear function of strike.

We also assume the local variance to be a piecewise linear function of strike. Moreover, for our model we obtained a closed form representation of the Put option prices via parameters of the model given in Sections 5.4, 5.5. Therefore, calibration of the model to the given set of smiles could be provided as follows. First, using the above-mentioned closed form solution for a fixed interval in x where parameters of the model are constant, we construct the combined solution valid for all $x \in \mathbb{R}^+$. At the second step, the parameters of the local variance function $v_{j,i}^0, v_{j,i}^1$ can be found together with the integration constants C_1, C_2 in Eq.(5.54) by solving a system of non-linear algebraic equations.

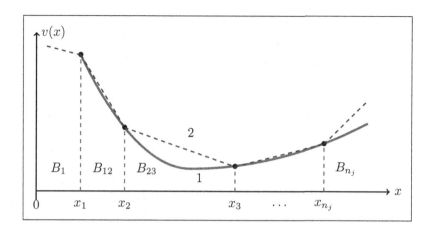

Figure 5.1: Construction of the combined solution in $x \in \mathbb{R}^+$: 1 (solid line — the real (unknown) local variance curve, 2 (dashed line) — a piecewise linear solution.

5.6.1 The combined solution in $x \in \mathbb{R}^+$

Suppose that the Put prices for $T = T_j$ are known for n_j ordered strikes. The location of these strikes on the x line is schematically depicted in Fig. 5.1.

Recall, that the Put prices are given by Eq.(5.54), which in a more convenient form at the interval $x_{i-1} \le x \le x_i$ and at $T = T_j$ can be represented as

$$P_i(x) = C_{j,i}^{(1)} \mathcal{J}_1(q_1, q_2, z) + C_{j,i}^{(2)} \mathcal{J}_2(q_1, q_2, z) + \frac{1}{b_1} I_{12}(z) + \mathcal{D}_j K, \quad (5.62)$$

$$z \equiv (b_2 + a_2 x) b_1 / a_2^2 = (v_{j,i}^0 + v_{j,i}^1 x) b_1 / a_2^2.$$

Here, for consistency we change notation of two integration constants which belong to the i-th interval in x and j-th maturity to $C_{j,i}^{(1)}, C_{j,i}^{(2)}$.

For the open interval B_1 in Fig. 5.1, since function $K_\nu(z)$ diverges when $z \to 0$, we have to put $C_{j,1}^{(1)} = 0$ as the boundary condition.[2] Therefore, Eq.(5.62) contains just one yet unknown constant $C_{j,1}^{(2)}$. For the closed intervals $x \in [x_{i-1}, x_i]$, $i \in [2, n_j]$ the solutions in Eq.(5.62) have two

---

[2]Actually, since $x \to 0$ implies $z = v \to b_2$, so b_2 should be non-negative, $b_2 \ge 0$. Therefore, the only case when $z \to 0$ at $x \to 0$ is when $b_2 = 0$.

yet unknown constants $C_{j,i}^{(1)}, C_{j,i}^{(2)}$, since x is finite on the corresponding intervals, and both solutions $y_1(x), y_2(x)$ are well-behaved. Finally, for the interval $x \in [x_{n_j}, \infty)$, according to the boundary conditions in Eq.(5.46) we must set $C_{2,n_j+1}^{(2)} = 0$.

Rigorously speaking, we also have to show that in the limits $x \to 0$ and $x \to \infty$ the source term $I_{12}(z)$ in Eq.(5.54) also vanishes. This could be done similar to Proposition 2 in [Itkin and Lipton (2018)].

Thus, we have $2n_j$ unknown constants to be determined. Since the local volatility function v_i is continuous at the points x_i, $i = 1, \ldots, n_j$, so should be the Put options prices $P(x, T_j)$. Therefore, we require that at the points x_i, $i = 1, ..., n_j$ the solution for Puts and its first derivative in x should be a continuous function of x. Thus, if the local variance function is known, the above constants solve a system of $2n_j$ algebraic equations. This system has a block-diagonal structure where each block is a 2×2 matrix. Therefore, it can be easily solved with the linear complexity $O(n_j)$.

When computing the first derivatives, we take into account that, [Abramowitz and Stegun (1964)]

$$\frac{\partial M(a,b,z)}{\partial z} = \frac{a}{b} M(a+1, b+1, z), \quad \frac{\partial U(a,b,z)}{\partial z} = -aU(a+1, b+1, z),$$

$$\partial_z I_{12}(z) = \left[\frac{y_1'(z)}{y_1(z)} I_1 + \frac{y_2'(z)}{y_2(z)} I_2 \right] a_2. \tag{5.63}$$

Therefore, computing the derivatives of the solution doesn't cause any new technical problem.

5.6.2 *Additional equations for calibration*

As we have already mentioned above, the standard way of doing calibration of the local volatility model would be that described, e.g., in [Itkin and Lipton (2018)]. Namely, given the maturity T_j and some initial guess of the local variance parameters $v_{j,i}^0, v_{j,i}^1, \forall i \in [1, n_j]$, the following steps represented in Panel 1 have to be achieved, e.g., in the standard least-square method.

Input: Strikes z_i, $i \in [1, n_j]$, Put prices V_i^{market}, $i \in [1, n_j]$
Output: $v_{j,i}^0, v_{j,i}^1, \forall i \in [1, n_j]$
Initialization: *The initial guess of* $v_{j,i}^0, v_{j,i}^1, \forall i \in [1, n_j]$, *the tolerance* ϵ;
while *1* **do**

    1. Solve the system for $C_{j,i}^{(1)}, C_{j,i}^{(2)}$;

    2. Compute Put option prices $V(x)$;

    3. Compute the total error $\Delta = \sum_{i=1}^{n_j} [V(x_i) - V^{market}(x_i)]^2$;

    if $\Delta > \epsilon$ **then**

       | New guess for $v_{j,i}^0, v_{j,i}^1, \forall i \in [1, n_j]$;

    else

       | break;

    end

end

Algorithm 1: Calibration of the local volatility model using a least-square method.

Here $V^{market}(z_i)$ are the market Put quotes at the given strikes and maturity. Obviously, when the number of calibration parameters (strikes) is high, this algorithm is slow even if the closed form solution is known and can be used at Step 2. Things become even worse when a numerical solution at Step 2 has to be used if the closed form solution is not available.

However, in our case this tedious algorithm can be fully eliminated. Indeed, at every point i in strike space, $i \in [1, n_j]$ we have four unknown variables $v_{j,i}^0, v_{j,i}^1, C_{j,i}^{(1)}, C_{j,i}^{(2)}$. We also have four equations which contain these variables, namely

$$P_i(x)|_{x=x_i} = P_{i+1}(x)|_{x=x_i}, \tag{5.64}$$

$$P_i(x)|_{x=x_i} = P_{market}(x_i),$$

$$\frac{\partial P_{i+1}(x)}{\partial x}\bigg|_{x=x_i} = \frac{\partial P_i(x)}{\partial x}\bigg|_{x=x_i},$$

$$v_{j,i}^0 + v_{j,i}^1 x_i = v_{j,i+1}^0 + v_{j,i+1}^1 x_i, \quad i = 1, \ldots, n_j.$$

Also, based on Eq.(5.51), the last line in Eq.(5.64) could be re-written as a recurrent expression

$$v_{j,i}^0 = v_{j,n_j}^0 + \sum_{k=i+1}^{n_j} x_k(v_{j,k}^1 - v_{j,k-1}^1), \quad i = 0, \ldots, n_j - 1. \tag{5.65}$$

The Eq.(5.64) is a system of $4n_j$ nonlinear equations with respect to $4(n_j + 1)$ variables $v_{j,i}^0$, $v_{j,i}^1$, $C_{j,i}^{(1)}$, $C_{j,i}^{(2)}$. We remind that according to the boundary conditions $C_{j,1}^{(1)} = C_{j,n_j}^{(2)} = 0$. Therefore, we need two additional conditions to provide a unique solution. For instance, often traders have an intuition about the asymptotic behavior of the volatility surface at infinity, which, according to our construction, is determined by v_{j,n_j}^1 and $v_{j,0}^1$.

Overall, solving the nonlinear system of equations Eq.(5.64) provides the final solution of our problem. This can be done by using standard methods, and, thus, no any optimization procedure is necessary. However, a good initial guess still would be helpful for a better (and faster) convergence.

5.6.3 *Smart initial guess*

The initial guess of the solution of Eq.(5.62) can be constructed, for instance, as follows. We take advantage of the fact that according to Eq.(5.49) the local variance function $v(x)$ could be explicitly expressed as

$$v(x) = \frac{b_1 x V_x(x) + b_0 V(x) - c}{V_{x,x}(x)}. \tag{5.66}$$

Given maturity T_j and approximating derivatives by central finite differences with the second order of approximation in step h in the strike space (see, e.g. [Itkin (2017)]), Eq.(5.66) can be represented in the form

$$v_{j,i}^0 + v_{j,i}^1 x_i = \frac{b_1 x V_x(x_i) + b_{0,j} V(x_i) - c_j}{V_{x,x}(x_i)}, \tag{5.67}$$

$$V_x(x_i) = \alpha_{-1} V(x_{i-1}) + \alpha_0 V(x_i) + \alpha_1 V(x_{i+1}),$$
$$V_{x,x}(x_i) = \delta_{-1} V(x_{i-1}) + \delta_0 V(x_i) + \delta_1 V(x_{i+1}),$$
$$\alpha_{-1} = -\frac{h_{i+1}}{h_i(h_{i+1} + h_i)}, \quad \alpha_0 = \frac{h_{i+1} - h_i}{h_{i+1} h_i}, \quad \alpha_1 = \frac{h_i}{h_{i+1}(h_{i+1} + h_i)}.$$
$$\delta_{-1} = \frac{2}{h_i(h_{i+1} + h_i)}, \quad \delta_0 = -\frac{2}{h_{i+1} h_i}, \quad \delta_1 = \frac{2}{h_{i+1}(h_{i+1} + h_i)}.$$
$$h_i = x_i - x_{i-1}, \quad i \in [1, n_j].$$

Further, associating Put prices $P(S, T_j, x_i)$ with the given market quotes, the right hands side of the first line in Eq.(5.67) can be found explicitly. This then can be combined with the last line of Eq.(5.64) to produce a system of $2(n_j - 1)$ equations for $v_{j,i}^1$ and $v_{j,i}^1$, $i \in [1, n_j]$. Finally, we take into account the asymptotic behavior of the volatility surface in x at zero and infinity, which, according to our construction, is determined by v_{j,n_j}^1 and $v_{j,0}^1$ and is assumed to be known. Thus, we obtain a closed system of $2(n_j - 1)$ linear equations with a banded matrix which can be easily solved with a linear complexity. This provides an explicit representation of the local variance function over the whole set of intervals in the strike space determined according to our approximation where the continuous derivatives are replace by finite differences.

Note, that at the first and last strike intervals the approximation of the first and second derivatives by central finite differences should be replaced by one-sided approximations, in more detail see [Itkin (2017)], chapter 2.

It could also happen that at some strikes this solution (the smart guess) gives rise to a negative local variance. In such a case we do another step which is a kind of smoothing. Namely, we exclude from the initial guess all values where the local variance is negative and using the remaining points create a spline. Then the negative values in the initial guess are replaced by those given by the constructed spline.

The final step utilizes the exact representation Eq.(5.62) of the Put price in the ELVG model. As the variance function is already known from the previous step, this equation contains two yet unknown constants $C_{j,i}^{(1)}, C_{j,i}^{(2)}$. Accordingly, they can be found by solving the system of 2 linear equations represented by the first and third lines of Eq.(5.64). Then, after this last step is complete, all unknown variables are determined, and thus found solution could be used as an educated initial guess for solving Eq.(5.64) numerically.

5.7 Asymptotic solutions

In many practical situations either some coefficients $a_2 = v_{j,i}^1$, or both $b_2 = v_{j,i}^0$, $a_2 = v_{j,i}^1$ in Eq.(5.52) are small. Of course, in that case the general solution Eq.(5.62) remains valid. However, in this case when computing the values of Kummer functions numerically, numerical errors significantly grow. This is especially pronounced when computing the integral I_{12}. The main point is that either the Kummer function takes a very small value, and then the constants $C_{j,i}^{(1)}, C_{j,i}^{(2)}$ should be big to compensate, or vice versa. Resolution of this requires a high-precision arithmetics, and, which is more important, taking many terms in a series representation of the Kummer functions, which significantly slows down the total performance of the method.

On the other hand, to eliminate these problems we can look for asymptotic solutions of Eq.(5.52) taking into account the existence of small parameters from the very beginning. This approach was successfully elaborated on in [Itkin and Lipton (2018)], and below we proceed in a similar spirit.

5.7.1 *Small a_2*

We can build the solution of Eq.(5.53) directly using an independent variable x (so not switching to the variable z). We represent it as a series on the small parameter a_2, i.e.

$$V(x) = \sum_{i=0}^{\infty} a_2^i V_i(x). \tag{5.68}$$

In the zero-order approximation by plugging Eq.(5.68) into Eq.(5.53) and neglecting by terms proportional to $a_2 \ll 1$ we obtain the following equation for $V_0(x)$

$$-b_2 V_{xx}(x) + b_1 x V_x(x) + b_0 V(x) = c. \tag{5.69}$$

This equation is simpler than Eq.(5.62). Still, its solution is given by a general formula

$$V(x) = C_1 y_1(x) + C_2 y_2(x) + I_{12}(x),$$

but the fundamental solutions $y_1(x), y_2(x)$ now read

$$y_1(x) = \mathcal{H}\left(-\frac{b_0}{b_1}, \sqrt{\frac{b_1}{2b_2}} x\right), \qquad y_2(x) = M\left(\frac{b_0}{2b_1}, \frac{1}{2}, \frac{b_1}{2b_2} x^2\right),$$

where $\mathcal{H}(a, x)$, $a, x \in \mathbb{R}$ is the generalized Hermite polynomial $H_a(x)$, [Abramowitz and Stegun (1964)].

5.7.2 *Small $|z|$*

Based on the definition of $z = (b_2 + a_2 x) b_1 / a_2^2$, this could occur in two cases: either at some finite interval in the strike space $|a_2| \gg |b_1 x|$, $|a_2| \gg |b_2|$, or just z is small, so b_2 and a_2 have the opposite signs. In any case we have a small parameter under the high-order derivative. This equation belongs to the class of singularly perturbed differential equations,

[Wasow (1987)]. It can be solved by using either the method of matching asymptotic expansions, [Nayfeh (2000)], or the method of boundary functions, [Vasil'eva et al. (1995)]. The latter was used in [Itkin and Lipton (2018)] in a similar situation, so for further details we refer a reader to that paper.

However, we can partly eliminate this by constructing solutions of Eq.(5.52) using the original variable x. Then we have to consider various cases where instead of a small parameter z some other combinations of parameters could be small or large. But if so, a general solution as a function of the original independent variable x could be represented as regular series on the new small parameter. Then, truncating the series, one gets a simplified solution.

To make it more transparent let us represent the general solution of Eq.(5.52) expressed in variable x, rather than in z, as this was done in Eq.(5.54)

$$V(x) = C_{j,i}^{(1)} y_1(x) + C_{j,i}^{(2)} y_2(x) + I_{12}(x), \tag{5.70}$$

$$y_i(x) = a_2^k (b_2 + a_2 x)^k \mathcal{V}_i \left(-1 - \frac{b_0}{b_1} + \frac{b_1 b_2}{a_2^2}, 2 - \frac{b_1 b_2}{a_2^2}, \frac{b_1}{a_2^2}(b_2 + a_2 x) \right),$$

$$i = 1, 2, \quad k = 1 - \frac{b_1 b_2}{a_2^2}.$$

Observe, that based on the definition of b_1 in Eq.(5.49), $b_1 \approx (r - q)\Delta T$, so usually small. Therefore, small z doesn't mean that w is necessarily small. Below we consider two cases.

5.7.2.1 $w \ll 1$

As $|z| \ll 1$ and $w \ll 1$ we have $w \ll |a_2^2/b_1|$. So $a_2 \geq \sqrt{b_1}$. In this case $w \ll 1$ is an actual small argument. Therefore, the general solution Eq.(5.70) can be expanded into series on small w. The condition $0 < w \ll 1$ implies that a_2 and b_2 have the opposite signs. If $a_2 > 0$ (and so $b_2 < 0$), then in the zero-order approximation we obtain

$$y_1(w) = (a_2 w)^{k-1} \left[\frac{\Gamma(-k)}{\Gamma(b_0/b_1)} a_2 w + O(w^2) \right] \tag{5.71}$$

$$- \left(\frac{b_1}{a_2^3} \right)^{1-k} \frac{\Gamma(k-1)}{b_1 \Gamma(k + b_0/b1)} (a_2 b_1 b_2 - b_0 a_2 w) + O(w^2),$$

$$y_2(w) = (a_2 w)^{k-1} \left[a_2 w + O(w^2) \right].$$

As $b_1 > 0$ we have $k - 1 > 0$.

If $a_2 < 0$ and $b_2 > 0$, then both RHS in Eq.(5.71) should be multiplied by a factor $\exp(-2i\pi b_1 b_2/a_2^2)$.

5.7.2.2 $a_2^2 \gg |b_1 w|$

In this case we can also expand the solution in Eq.(5.70) into series on small z to obtain

$$y_1(z) = \frac{1}{\Gamma(1 + q_1 - q_2)} \left[\Gamma(1 - q_2) - q_1 \Gamma(-q_2)z \right] \qquad (5.72)$$

$$+ z^{-q_2} \left[\frac{\Gamma(q_2 - 1)}{\Gamma(q_1)} z + O(z^2) \right] + O(z^2),$$

$$y_2(w) = 1 + \frac{q_1}{q_2} z + O(z^2).$$

Note, that based on the definition $q_2 = b_2 b_1/a_2^2$, at large a_2 the coefficient q_2 could also be small. But $z/q_2 = 1 + a_2 x/b_2 = w/b_2 = O(1)$.

5.8   Numerical experiments

In our numerical test we use the same data set as in [Itkin (2015); Itkin and Lipton (2018)]. This is done first, to compare performance and a quality of the fit for all those models. Also, we already know that these smiles are difficult to fit precisely, see discussions in [Itkin (2015); Itkin and Lipton (2018)].

To remind, we take data from http://www.optionseducation.org on XLF traded at NYSEArca on March 25, 2014. The spot price of the index is $S = 22.64$, and $r = 0.0148$, $q = 0.01$. The option implied volatilities (I_{call}, I_{put}) are given in Tables 5.1,5.2. We take all OTM quotes and some ITM quotes which are very close to the at-the-money (ATM). When strikes for Calls and Puts coincide, we take an average of I_{call} and I_{put} with weights proportional to $1 - |\Delta|_c$ and $1 - |\Delta|_p$ respectively, where Δ_c, Δ_p are option Call and Put deltas.[3]

We have already mentioned that in our model for each term the slopes of the smile at plus and minus infinity, v_{j,n_j}^1 and $v_{j,0}^1$, are free parameters. So often traders have an intuition about these values. However, in our numerical experiments we take for them just some plausible values. In more

---

[3]By doing so we do take into account effects reported in [Ahoniemi (2009)], who pointed out that the IVs calculated from Call and Put option prices corresponding to the same strike do not coincide, although they should be equal in theory. Our weights are chosen according to a pure empirical rule of thumb, and a more detailed investigation of this effect is required.

Table 5.1: XLF implied volatilities for the Put options.

T	10	11	12	13	14	15	16	17	18	19	19.5	20	20.5	21	21.5	22	23
												K,Put					
4/4/2014												39.53		23.77	19.73	16.67	
4/11/2014											35.89	30.33	26.62	22.06	18.49	16.11	
4/19/2014										32.90		26.79		20.14		15.19	12.93
5/17/2014								37.66	33.27	26.88		23.08		18.94		16.12	13.86
6/21/2014						40.51	37.21	31.41	27.84	23.90		21.07		18.88		16.95	15.82
7/19/2014						36.71	33.35	29.96	26.09	22.81		20.29		18.13		16.30	14.93
12/20/2014					31.98	29.38	27.21	25.30	23.75	22.09		20.67		19.44		18.36	17.60
1/17/2015	42.75	38.79	35.60	33.26	30.94	28.82	26.52	24.96	23.12	21.67		20.29		19.10		17.90	18.07

Table 5.2: XLF implied volatilities for the Put options.

T	21	21.5	22	22.5	23	23.5	24	25	26	27	28	29	30
							K,Call						
4/4/2014		16.60	14.69	14.40	14.86								
4/11/2014		16.89	14.96	14.52	14.77	14.98							
4/19/2014			15.79		13.38		15.39						
5/17/2014	16.71		14.48				13.75						
6/21/2014	16.31		14.78				13.92	14.28	16.58				
7/19/2014	16.82		15.24				14.36	14.19	15.20				
12/20/2014	17.63		16.61				15.86	15.47	15.12	15.18	15.03		
1/17/2015					16.95		17.25	16.23	15.73	15.50	15.58	15.86	16.47

detail, for a normalized variance $v(x)$ defined in Eq.(5.49), for all smiles we use $v_{j,0}^1 = -0.1$, and $v_{j,n_j}^1 = 0.1$. Accordingly, for the instantaneous variance $\sigma^2(x) = 2S^2 v(x)/p_j$ the slopes at both zero and plus infinity are time-dependent and can be computed by using the above formula.

When calibrating the model to market data, we use the standard Matlab *fsolve* function, and utilize a "trust-region-dogleg" algorithm (see Matlab documentation on *fsolve*). Parameter "TypicalX" has to be chosen carefully to speedup calculations.

The results of this calibration which is done term-by term, are given in Fig. 5.2. Here each subplot corresponds to a single maturity T (marked in the legend) and shows market data (discrete points) and computed values (solid line). It can be seen that this simple local calibration algorithm provides a very accurate fit for all terms.[4]

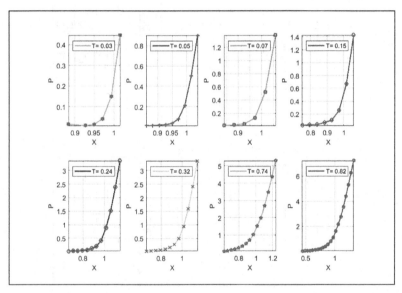

Figure 5.2: Term-by-term fitting of market Put prices constructed using the whole set of data in Tab. 5.1,5.2.

We constructed the calibration algorithm to be smart enough in a sense that based on the values of parameters at each iteration it decides itself which particular solution (full or asymptotic) should be used at this iter-

---

[4]Note, that in [Itkin and Lipton (2018)] in the last subplot the fit is not perfect in the vicinity of $X = -0.5$, where $X = \log K/F$ and $F = Se^{(r-q)T}$.

ation. We also observed that all full and asymptotic solutions are utilized by the algorithm when calibrating these market smiles.

Table 5.3 presents some performance measures of our algorithm. It can be seen that the elapsed time depends on the number of iterations and function evaluations necessary to converge to the given tolerance (we use a relative tolerance $\varepsilon = 10^{-4}$). This, in turn, depends on the number of evaluated Kummer functions (for the full solution), or number of exponential and Gamma functions (for the asymptotic solutions). Of course, the asymptotic solutions are much faster to evaluate, therefore an average time to calibrate a typical term is less than a second. For the last term 8 in Tab. 5.3 calibration is slow for two reasons: (i) full solution is used based on the values of parameters, and (ii) the number of strikes is higher than for the other terms. But the main reason is that the market data for this term is quite irregular. In any case, performance of this model is much better than that reported in both [Itkin (2015)] and [Itkin and Lipton (2018)].

Table 5.3: Performance characteristics of the algorithm in the described experiment.

Term	T, years	Elapsed time, sec	iterations	function evaluations	strikes
1	0.0274	0.86	97	1202	6
2	0.0466	2.83	97	1808	9
3	0.0685	1.43	95	1200	6
4	0.1452	0.64	48	433	8
5	0.2411	0.90	37	470	12
6	0.3178	2.98	82	1523	12
7	0.7397	6.60	106	3017	15
8	0.8164	149.67	56	1317	21

The local variance curves obtained as a result of this fitting are given term-by-term in Fig. 5.3. The corresponding local variance surface is represented in Fig. 5.4

By comparing the surface with that given in [Itkin and Lipton (2018)], one can notice that the shape is quite different while for calibration we use the same market smiles. This is because in [Itkin and Lipton (2018)] the standard local volatility model is used, where the underlying price follows a Geometric Brownian motion equipped with an instantaneous local volatility function, while here the model is quite different.

To look at a more regular surface, we proceed with another example which is taken from [Balaraman (2016)]. In that paper an implied volatility

Fitting Local Volatility

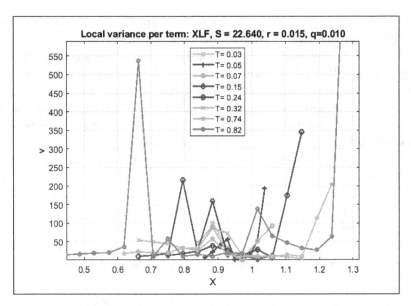

Figure 5.3: Term-by-term fitting of the instantaneous local variance $\sigma^2(x,T)$.

surface of S&P500 is presented, and the local volatility surface is constructed using the Dupire formula. In our test we take data for the first 12 maturities and all strikes as they are given in [Balaraman (2016)], and apply our model to calibrate the local variance surface as this is described in above. When doing so we set $v^1_{j,0} = -0.3$, and $v^1_{j,n_j} = 0.1$ for all smiles.

The results of this calibration are presented in Figs. 5.5,5.6,5.7. By construction, our surface preserves no-arbitrage, while for the approach in [Balaraman (2016)] they have to solve some additional problems.[5]

In Table 5.4 we present the performance of our algorithm in this experiment. It can be seen that here the elapsed time is similar or shorter as compared with the previous test presented in Table 5.3.

---

[5]As this is mentioned in [Balaraman (2016)], the correct pricing of local volatility surface requires an arbitrage free implied volatility surface. If the input implied volatility surface is not arbitrage free, this can lead to negative transition probabilities and/or negative local volatilities and can give rise to mispricing.

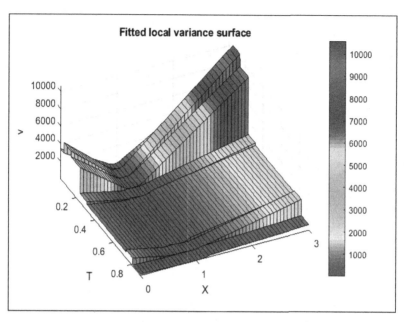

Figure 5.4: The instantaneous local variance surface $\sigma^2(x, T)$ constructed by using the proposed approach.

Table 5.4: Performance characteristics of the algorithm for calibration of a S&P500 surface.

Term	T, years	Elapsed time, sec	iterations	function evaluations	strikes
1	0.0822	1.09	99	1604	8
2	0.1671	0.56	40	377	8
3	0.2521	2.32	94	1615	8
4	0.3315	1.70	97	1186	8
5	0.4164	0.10	15	64	8
6	0.4986	2.35	111	1600	8
7	0.5836	2.40	111	1584	8
8	0.6658	2.25	131	1604	8
9	0.7507	1.51	95	1072	8
10	0.8356	2.30	98	1603	8
11	0.9178	0.07	13	46	8
12	1.0027	72.80	74	795	8

At the end, it is worth to underline that the ELVG model relies on a no-arbitrage interpolation, and then constructs a closed-form solution of the

Fitting Local Volatility

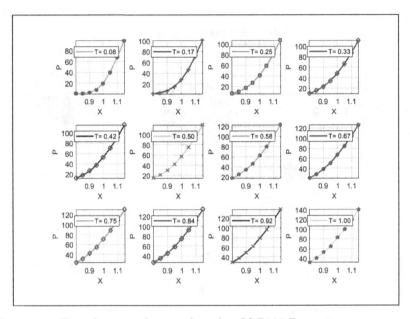

Figure 5.5: Term-by-term fitting of market S&P500 Put prices constructed using data of [Balaraman (2016)].

Dupire-wise ODE in terms of hypergeometric and generalized hypergeometric functions. An important advantage of this approach is that calibration of the model to market smiles does not require solving any optimization problem, and can be done term-by-term by solving a system of non-linear algebraic equations for each maturity, which, in general, is significantly faster, especially since we provide an algorithm for constructing a smart initial guess. Also, we provide various asymptotic solutions which allow a significant acceleration of the numerical solution and improvement of its accuracy in the corresponding cases (i.e, when parameters of the model at some iteration obey the conditions to apply the corresponding asymptotic).

In principle, somebody could claim that solving a system of nonlinear equations with a generic solver is not much different from solving a nonlinear optimization problem. Obviously, when our ODE is used as an alternative to the Dupire equation, the difference comes from the fact that calibration based on the Dupire equation requires solving this PDE at every iteration by either numerically, or semi-analytically by using a Laplace transform, which is obviously slower. As was mentioned in Introduction there exist

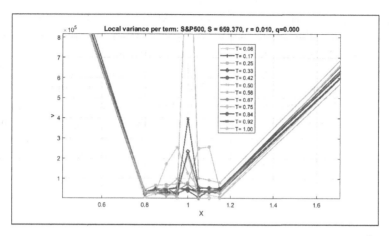

Figure 5.6: Term-by-term fitting of the instantaneous local variance $\sigma^2(x, T)$ for S&P500.

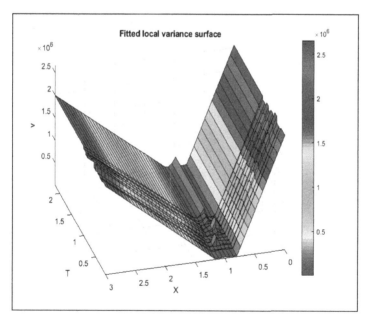

Figure 5.7: The instantaneous local variance surface $\sigma^2(x, T)$ for S&P500 constructed by using the proposed approach.

many other calibration algorithms which reduce to a nonlinear optimization problem (e.g., taking a sufficiently large parametric family of local volatility functions and choosing the parameters that provide the best fit of observed prices). For the latter computation of the objective function is fast, but optimization must be constrained to preserve no-arbitrage, and, thus, slow.

The results of the test demonstrate robustness of the proposed approach from both the speed and accuracy point of view, especially in cases where the above referred papers experienced some difficulties with achieving a perfect fit. An additional test performed for the S&P500 data taken from [Balaraman (2016)] gives rise to the same conclusion.

Chapter 6

Geometric Local Variance Gamma Model

As mentioned in Chapter 5, the Local Variance Gamma (LVG) volatility model was first introduced by P. Carr in 2008 and then presented in [Carr and Nadtochiy (2014, 2017)] as an extension of the local volatility model by [Dupire (1994)] and [Derman and Kani (1994a)]. The latter was developed on top of the celebrating Black-Scholes model to take into account the existence of option smile. The main advantage of all local volatility models is that given European options prices or their implied volatilities at points (T, K) where K, T are the option strike and time to maturity, they are able to exactly replicate the local volatility function $\sigma(T, K)$ at these points. This process is called calibration of the local volatility (or, alternatively, implied volatility) surface, and is one of the main topics of this book.

In Chapter 5 we described an extension of the Local Variance Gamma model which is called ELVG and was originally proposed in [Carr and Itkin (2018a)]. As compared with the classical local volatility model, the LVG and ELVG have several advantages. First, they are richer in the financial sense. Indeed, it is worth noting that the term "local" in the name of the LVG/ELVG models is a bit confusing. This is because, e.g., the ELVG is constructed by equipping an arithmetic Brownian motion with drift and local volatility by stochastic time change $\Gamma_{X(t)}$. Here Γ_t is a Gamma stochastic variable, and $X(t)$ is a deterministic function of time t. As stochastic change is one of the ways of introducing stochastic volatility, it could be observed that the LVG/ELVG is actually a local stochastic volatility (LSV) model which combines local and stochastic features of the volatility process. For more information on the LSV models, see [Bergomi (2016); Kienitz and Wetterau (2012)].

Another advantage of the LVG/ELVG is that their calibration is computationally more efficient. This is because this construction gives rise not to a partial differential equation (which in the classical case is known as Dupire's equation), but to a partial differential difference equation (PDDE). The latter is actually an ordinary differential equation (ODE) and permits both explicit calibration and fast numerical valuation. In particular, calibration of the local variance surface does not require any optimization method, rather just a root solver, [Carr and Itkin (2018a)].

As discussed in [Itkin and Lipton (2018)], given the market quotes of European options for various maturities and strikes, the local (and then implied) volatility surface can be obtained by directly solving the Dupire equation using either analytical or numerical methods. The advantage of such an approach is that it guarantees no-arbitrage if the corresponding analytical or numerical method does preserve no-arbitrage (including various interpolations, etc.). Obviously, solving Dupire's PDE requires either numerical methods, e.g. that in [Coleman *et al.* (2001)], or, as in [Itkin and Lipton (2018)], a semi-analytic method which: (i) first uses the Laplace-Carson transform, and (ii) then applies various transformations to obtain a closed form solution of the transformed equation in terms of Kummer Hypergeometric functions. Still, it requires an inverse Laplace transform to obtain the final solution. To make the second approach tractable, some assumptions should be made about the behavior of the local/implied volatility surface at strikes and maturities where the market quotes are not known. Usually, the corresponding local variance is assumed to be either piecewise constant, [Lipton and Sepp (2011a)], or piecewise linear [Itkin and Lipton (2018)] in the log-strike space, and piecewise constant in the time to maturity space. A similar assumption is also necessary to make the LVG/ELVG models tractable. In particular, in [Carr and Nadtochiy (2017)] the model was calibrated to option smiles assuming the local variance is a *piecewise constant* function of strike, while in [Carr and Itkin (2018a)] the local variance is a *piecewise linear* function of strike.

Despite these nice features of the ELVG, one possible problem could be that the model is developed based on the arithmetic Brownian motion with drift. That means that the underlying, in principle, could acquire negative values, which in some cases is undesirable, e.g., if the underlying is a stock price. Therefore, another extension of the LVG model was developed in [Carr and Itkin (2018b)] which operates with a Gamma time-changed *geometric* Brownian motion with drift, and the local variance which is a function of the spot level only (so is not a function of time).

Second, in [Carr and Nadtochiy (2017)] the model was calibrated to option smiles assuming the local variance is a *piecewise constant* function of strike, while in [Carr and Itkin (2018a)] the local variance is a *piecewise linear* function of strike. In [Carr and Itkin (2018b)] we consider three *piecewise linear* models: the local variance as a function of strike, the local variance as a function of log-strike, and the local volatility as a function of strike (so, the local variance is a *piecewise quadratic* function of strike). We show that in this new model it is still possible to derive an ordinary differential equation for the option price, which plays the role of Dupire's equation for the standard local volatility model. Moreover, it all three cases, this equation can be solved in closed form. Finally, similar to [Carr and Itkin (2018a)] we show that given multiple smiles the whole local variance/volatility surface can be recovered which does not require solving any optimization problem. Instead, it can be done term-by-term, and for every maturity the entire calibration is done by solving a system of non-linear algebraic equations which is significantly faster.

In this Chapter we describe in detail the approach of [Carr and Itkin (2018b)] in detail. The Chapter has the following structure. In Section 6.1 the new model, which for an obvious reason we call the Geometric Local Variance Gamma model or the GLVG, is formulated. In Section 6.2 we derive a forward equation (which is an ordinary differential equation (ODE)) for Put option prices using a homogeneous Bochner subordination approach. Section 6.3 generalizes this approach by considering the local variance being piecewise constant in time. A closed form solution of the derived ODE is given in terms of Hypergeometric functions for various models of the local variance or volatility. The next Section discusses computation of a source term of this ODE which requires a no-arbitrage interpolation. Using the idea of [Itkin and Lipton (2018)]), we show how to construct non-linear interpolation which provides both no-arbitrage, and a nice tractable representation of the source term, so that all integrals in the source term can be computed in closed form. In Section 6.5 calibration of multiple smiles in our model is discussed in detail. To calibrate a single smile we derive a system of nonlinear algebraic equations for the model parameters, and explain how to obtain a smart guess for their initial values. In Section 6.6 we discuss the results of some numerical experiments where calibration of the model to the given market smiles is done term-by-term.

Despite some similarity with the ELVG model, Section 5.1, the derivation of the time-changed function is done in a different (reverse engineering) way which deserves a full description. This is discussed in Section 6.1.

6.1 Stochastic model

Let W_t be a $\mathbb{Q}$ standard Brownian motion with time index $t \geq 0$. Consider a stochastic process D_t to be a time-homogeneous diffusion with drift μ

$$dD_t = \mu D_t dt + \sigma(D_t) D_t dW_t, \qquad (6.1)$$

where the volatility function σ is local and time-homogeneous.

A unique solution to Eq.(6.1) exists if $\sigma(D) : \mathbb{R} \to \mathbb{R}$ is Lipschitz continuous in D and satisfies growth conditions at infinity. Since D is a time-homogeneous Markov process, its infinitesimal generator $\mathcal{A}$ is given by

$$\mathcal{A}\phi(D) \equiv \left[\mu D \nabla_D + \frac{1}{2}\sigma^2(D) D^2 \nabla_D^2 \right] \phi(D), \qquad (6.2)$$

for all twice differentiable functions ϕ. Here ∇_x is a first order differential operator (the first derivative) on x. The semigroup of the D process (which here is an expectation under $\mathbb{Q}$) is

$$\mathcal{T}_t^D \phi(D_t) = e^{t\mathcal{A}}\phi(D_t) = \mathbb{E}_{\mathbb{Q}}[\phi(D_t)|D_0 = D], \quad \forall t \geq 0. \qquad (6.3)$$

This first equality could be also thought of as the Feynman-Kac theorem representation of the solution to the terminal value problem (see, e.g., [Lörinczi *et al.* (2011)]), which connects the expectation in the right hand side to the solution of the corresponding PDE, and then the formal solution of this PDE is given by the exponential operator $e^{t\mathcal{A}}$ applied to the initial condition $\phi(D_t)$.

In the spirit of [Carr and Nadtochiy (2017); Carr and Itkin (2018a)], introduce a new process D_{Γ_t} which is D_t subordinated by the unbiased Gamma clock Γ_t. The density of the unbiased Gamma clock Γ_t at time $t \geq 0$ is

$$\mathbb{Q}\{\Gamma_t \in d\nu\} = \frac{\nu^{m-1} e^{-\nu m/t}}{(t^*)^m \Gamma(m)} d\nu, \quad \nu > 0, \quad m \equiv t/t^*. \qquad (6.4)$$

Here $t^* > 0$ is a free parameter of the process, $\Gamma(x)$ is the Gamma function. It is easy to check that

$$\mathbb{E}_{\mathbb{Q}}[\Gamma_t] = t. \qquad (6.5)$$

Thus, on average the stochastic gamma clock Γ_t runs synchronously with the calendar time t.

As applied to the option pricing problem, we introduce a more complex construction. Namely, consider options written on the underlying process

S_t. Without loss of generality and for the sake of clearness let us treat below S_t as the stock price process. Let us define S_t as

$$S_t = D_{\Gamma_{X(t)}} \tag{6.6}$$

where $X(t)$ is a deterministic function of time t. We need to determine $X(t)$ such that under a risk-neutral measure $\mathbb{Q}$, the total gains process $\hat{S}_t$, including the underlying price appreciation and continuous dividends q, after discounting at the risk free rate r is a martingale, see [Shreve (1992)].

Taking first a derivative of $\hat{S}_t$

$$d\hat{S}_t = d\left(e^{-rt}S_t e^{qt}\right) = e^{(q-r)t}\left[(q-r)S_t dt + dS_t\right], \tag{6.7}$$

and then an expectation of both parts we obtain

$$\mathbb{E}_{\mathbb{Q}}[d\left(e^{(q-r)t}S_t\right)] = e^{(q-r)t}\left\{(q-r)\mathbb{E}_{\mathbb{Q}}[S_t]dt + d\mathbb{E}_{\mathbb{Q}}[S_t]\right\}. \tag{6.8}$$

So in order for $\hat{S}_t$ to be a martingale, the RHS of Eq.(6.8) should vanish. Solving the equation

$$(q-r)y(t)dt + dy(t) = 0, \qquad y(t) = \mathbb{E}_{\mathbb{Q}}[S_t|S_s], \ s < t,$$

we obtain

$$y(t) = \mathbb{E}_{\mathbb{Q}}[S_t|S_s] = S_s e^{(r-q)(t-s)}, \tag{6.9}$$

$$\mathbb{E}_{\mathbb{Q}}[dS_t|S_s] = d\mathbb{E}_{\mathbb{Q}}[S_t|S_s] = S_s(r-q)e^{(r-q)(t-s)}dt.$$

On the other hand, from Eq.(6.6)

$$\mathbb{E}_{\mathbb{Q}}[dS_t|S_s] = \mathbb{E}_{\mathbb{Q}}[dD_{\Gamma_{X(t)}}|S_s] = \mu\mathbb{E}_{\mathbb{Q}}[D_{\Gamma_{X(t)}}d\Gamma_{X(t)}|S_s] \tag{6.10}$$

$$+ \mathbb{E}_{\mathbb{Q}}[\sigma(D_{\Gamma_{X(t)}})D_{\Gamma_{X(t)}}dW_{\Gamma_{X(t)}}|S_s] = \mu\mathbb{E}_{\mathbb{Q}}[D_{\Gamma_{X(t)}}d\Gamma_{X(t)}|S_s],$$

because the process $W_{\Gamma_{X(t)}}$ is a local martingale, see [Revuz and Yor (1999)], Chapter 6. Accordingly, the process $W_{\Gamma_{X(t)}}$ inherits this property from W_{Γ_t}, hence $\mathbb{E}_{\mathbb{Q}}[\sigma(D_{\Gamma_{X(t)}})D_{\Gamma_{X(t)}}dW_{\Gamma_{X(t)}}] = 0$.

To proceed, assume the Gamma process Γ_t is independent of W_t (and, accordingly, $\Gamma_{X(t)}$ is independent of $W_{\Gamma_{X(t)}}$). Then the expectation in the RHS of Eq.(6.10) can be computed, by first conditioning on $\Gamma_{X(t)}$, and then integrating over the distribution of $\Gamma_{X(t)}$ which can be obtained from Eq.(6.4) by replacing t with $X(t)$, i.e.

$$\mathbb{E}_{\mathbb{Q}}[D_{\Gamma_{X(t)}}d\Gamma_{X(t)}|S_s] = \int_0^\infty \mathbb{E}_{\mathbb{Q}}[D_{\Gamma_{X(t)}}d\Gamma_{X(t)}|\Gamma_{X(t)} = \nu]\frac{\nu^{m-1}e^{-\nu m/X(t)}}{(t^*)^m\Gamma(m)} \tag{6.11}$$

$$= \int_0^\infty \mathbb{E}_{\mathbb{Q}}[D_\nu]\frac{\nu^{m-1}e^{-\nu m/X(t)}}{(t^*)^m\Gamma(m)}d\nu,$$

$$\nu > 0, \quad m \equiv X(t)/t^*.$$

To find $\mathbb{E}_{\mathbb{Q}}[D_\nu]$ we take into account Eq.(6.1) to obtain

$$d\mathbb{E}_{\mathbb{Q}}[D_\nu] = \mathbb{E}_{\mathbb{Q}}[dD_\nu] = \mathbb{E}_{\mathbb{Q}}[\mu D_\nu d\nu + \sigma(D_\nu)D_\nu dW_\nu] = \mu \mathbb{E}_{\mathbb{Q}}[D_\nu]d\nu. \quad (6.12)$$

Solving this equation with respect to $y(\nu) = \mathbb{E}_{\mathbb{Q}}[D_\nu|D_s]$, we obtain $\mathbb{E}_{\mathbb{Q}}[D_\nu|D_s] = D_s e^{\mu(\nu-s)}$. Since we condition on time s, it means that $D_s = D_{\Gamma_{X(s)}} = S_s$, and thus $\mathbb{E}_{\mathbb{Q}}[D_\nu|D_s] = S_s e^{\mu(\nu-s)}$.

Further, we substitute this into Eq.(6.11), set the parameter of the Gamma distribution t^* to be $t^* = X(t)$ (so $m = 1$) and integrate to obtain

$$\mathbb{E}_{\mathbb{Q}}[dS_t|S_s] = \mu \mathbb{E}_{\mathbb{Q}}[D_{\Gamma_{X(t)}} d\Gamma_{X(t)}] = S_s e^{-s\mu} \frac{\mu}{1 - \mu X(t)} dt. \quad (6.13)$$

Finally, equating representations of $\mathbb{E}_{\mathbb{Q}}[dS_t|S_s]$ obtained in Eq.(6.9) and Eq.(6.13) we arrive at the equation for $X(t)$

$$S_0(r - q)e^{(r-q)(t-s)} = S_s e^{-s\mu} \frac{\mu}{1 - \mu X(t)}. \quad (6.14)$$

Assuming $\mu = r - q$, this equation can be solved to provide

$$X(t) = \frac{1 - e^{-(r-q)t}}{r - q}. \quad (6.15)$$

A similar expression for $X(t)$ was also used in [Carr and Itkin (2018a)] for the ELVG. We already mentioned that the ELVG could be considered as an *arithmetic analog* of the described model, which is *geometric* in D_t.

It is clear that in the limit $r \to 0$, $q \to 0$ we have $X(t) = t$. Also based on Eq.(6.5)

$$\mathbb{E}_{\mathbb{Q}}[\Gamma_{X(t)}] = X(t). \quad (6.16)$$

The function $X(t)$ starts at zero, i.e. $X(0) = 0$,[1] and is a continuous non-decreasing function of time t. In more detail, if $r - q > 0$, the function $X(t)$ is increasing in t in all points except at $t \to \infty$, where it tends to constant. However, the infinite time horizon doesn't have much practical sense, therefore for any finite time t the function $X(t)$ can be treated as an increasing function in t. In the other case when $r - q < 0$, the function $X(t)$ is strictly increasing $\forall t \in [0, \infty)$. This means that, overall, $X(t)$ has all properties of a good clock. Accordingly, $\Gamma_{X(t)}$ has all properties of a random time.

Thus, we managed to demonstrate that with this choice of μ and $X(t)$ the right hands part of Eq.(6.8) vanishes, and our discounted stock process with allowance for non-zero interest rates and continuous dividends becomes a martingale. So the proposed construction can be used for option pricing.

---

[1]So our assumption made in above that $X(0) = 0$ is consistent.

This setting can be easily generalized for time-dependent interest rates $r(t)$ and continuous dividends $q(t)$. We leave it for the reader.

The next step is to establish a connection between the original and time-changed processes. It is known from [Bochner (1949)] that the process G_{Γ_t} defined as

$$dG_t = \sigma^2(G)G_t dW_t,$$

is a time-homogeneous Markov process. Same is true for the process $(r - q)G_t dt$. Thus, the entire process D_t defined in Eq.(6.1) is also a time-homogeneous Markov process. Accordingly, the semigroups T_t^S of S_t and T_t^D of $D_{\Gamma_{X(t)}}$ are connected by the Bochner integral[2]

$$T_t^S U(S) = \int_0^\infty T_\nu^D U(S) \mathbb{Q}\{\Gamma_{X(t)} \in d\nu\}, \quad \forall t \geq 0, \tag{6.17}$$

where $U(S)$ is a function in the domain of both T_t^D and T_t^S. It can be derived by exploiting the time homogeneity of the D process, conditioning on the gamma time first, and taking into account the independence of Γ_t and W_t (or $\Gamma_{\Gamma_{X(t)}}$ and $W_{\Gamma_{X(t)}}$ in our case).

As we set parameter t^* of the gamma clock to $t^* = X(t)$, Eq.(6.17) and Eq.(6.4) imply

$$T_t^S U(S) = \int_0^\infty T_\nu^D U(S) \frac{e^{-\nu/X(t)}}{X(t)} d\nu. \tag{6.18}$$

In what follows for the sake of brevity we call this model as the Geometric Local Variance Gamma model, or the GLVG.

6.2 Forward equation for option prices

In this section we derive a forward equation for put option prices, which is an analog of the Dupire equation for the standard local volatility model. In doing so, we closely follow the description in the corresponding section of [Carr and Itkin (2018a)], as from the derivation point of view the GLVG differs from the ELVG just by the definitions of infinitesimal generator $\mathcal{A}$ of the process D_t.

Let us interpret the index t of the semigroup T_t^S as the maturity date T of a European claim with the valuation time $t_v = 0$. Also let the test function $U(S)$ be the payoff of this European claim, i.e.

$$U(S_T) = e^{-rT}(K - S_T)^+. \tag{6.19}$$

---

[2]Here it represents an expectation of the option price with respect to the second stochastic driver — stochastic clock ν.

Then define

$$P(S_0, T, K) = \mathcal{T}_T^S U(S_0), \tag{6.20}$$

as the European Put value with maturity T at time $t = 0$ in the ELVG model. Similarly

$$P^D(S_0, \nu, K) = \mathcal{T}_\nu^D U(S_0), \tag{6.21}$$

would be the European Put value with maturity ν at time $t = 0$ in the model of Eq.(6.1).[3] Then the Bochner integral in Eq.(6.18) takes the form

$$P(S, T, K) = \int_0^\infty P^D(S, \nu, K) p e^{-p\nu} d\nu, \quad p \equiv 1/X(T). \tag{6.22}$$

Thus, $P(S, T, K)$ is represented by a Laplace-Carson transform of $P^D(S, \nu, K)$ with p being the transform parameter. Note that

$$P(S, 0, K) = P^D(S, 0, K) = U(S). \tag{6.23}$$

To proceed, we need an analog of the Dupire forward PDE for $P^D(S, \nu, K)$.

6.2.1 *Dupire-like forward PDE*

The derivation of the Dupire-like forward PDE is similar to that given in Section 5.2.1. Nevertheless, here we provide it in full for completeness.

First, differentiating Eq.(6.21) by ν with allowance for Eq.(6.3) yields

$$\nabla_\nu P^D(S, \nu, K) = e^{-r\nu} e^{\nu \mathcal{A}} [\mathcal{A} - r] U(S) = e^{-r\nu} \mathbb{E}_{\mathbb{Q}} [\mathcal{A} - r] U(S). \tag{6.24}$$

We take into account the definition of the generator $\mathcal{A}$ in Eq.(6.2), and also remind that at $t = 0$ we have $D_0 = S_0 \equiv S$. Then Eq.(6.24) transforms to

$$\nabla_\nu P^D(S, \nu, K) = -r P^D(S, \nu, K) + (r - q) S \nabla_S P^D(S, \nu, K) \tag{6.25}$$

$$+ e^{-r\nu} \frac{1}{2} \mathbb{E}_{\mathbb{Q}} \left[\sigma^2(S) S^2 \nabla_S^2 U(S) \right].$$

However, we need to express the forward equation using a pair of independent variables (ν, K) while Eq.(6.24) is derived in terms of (ν, S). To do this, observe that

$$\mathbb{E}_{\mathbb{Q}} \left[\sigma^2(S) S^2 \nabla_S^2 U(S) \right] = \mathbb{E}_{\mathbb{Q}} \left[\sigma^2(S) S^2 \delta(K - S) \right] = \mathbb{E}_{\mathbb{Q}} \left[\sigma^2(K) K^2 \delta(K - S) \right]$$

$$= \mathbb{E}_{\mathbb{Q}} \left[\sigma^2(K) K^2 \nabla_K^2 U(S) \right] = e^{r\nu} \sigma^2(K) \nabla_K^2 P^D(S, \nu, K). \tag{6.26}$$

---

[3]Below for simplicity of notation we drop the subscript '0' in S_0.

where the sifting property of the Dirac delta function $\delta(S - K)$ has been used. Also

$$-rP^D(S, \nu, K) + (r - q)S\nabla_S P^D(S, \nu, K) \qquad (6.27)$$

$$= e^{-r\nu}\mathbb{E}_\mathbb{Q}\left[-r(K - S)^+ + (r - q)S\frac{\partial(K - S)^+}{\partial S}\right]$$

$$= e^{-r\nu}\mathbb{E}_\mathbb{Q}\left[-r(K - S)^+ - (r - q)(K - S)\frac{\partial(K - S)^+}{\partial S}\right.$$

$$\left. + (r - q)K\frac{\partial(K - S)^+}{\partial S}\right]$$

$$= e^{-r\nu}\mathbb{E}_\mathbb{Q}\left[-r(K - S)^+ + (r - q)(K - S)^+ - (r - q)K\frac{\partial(K - S)^+}{\partial K}\right]$$

$$= -qP^D(S, \nu, K) - (r - q)K\nabla_K P^D(S, \nu, K).$$

Therefore, using Eq.(6.26) and Eq.(6.27), Eq.(6.24) could be transformed to

$$\nabla_\nu P^D(S, \nu, K) = -qP^D(S, \nu, K) - (r - q)K\nabla_K P^D(S, \nu, K) \qquad (6.28)$$

$$+ \frac{1}{2}\sigma^2(K)K^2\nabla_K^2 P^D(S, \nu, K) \equiv \mathcal{A}^K P^D(S, \nu, K),$$

$$\mathcal{A}^K = -q - (r - q)K\nabla_K + \frac{1}{2}\sigma^2(K)K^2\nabla_K^2.$$

This equation looks exactly like the Dupire equation with non-zero interest rates and continuous dividends, see, e.g., [Ekström and Tysk (2012)] and references therein. Note, that $\mathcal{A}^K$ is also a time-homogeneous generator.

6.2.2 *PDDE for a single term*

Our final step is to apply the linear differential operator $\mathcal{L}$ defined in Eq.(6.28) to both parts of Eq.(6.22). Using time-homogeneity of D_t and again the Dupire equation Eq.(6.28), we obtain

$$-qP(S, T, K) - (r - q)K\nabla_K P(S, T, K) + \frac{1}{2}\sigma^2(K)K^2\nabla_K^2 P(S, T, K)$$

$$= \int_0^\infty pe^{-p\nu}\left[-qP^D(S, \nu, K) - (r - q)K\nabla_K P^D(S, \nu, K)\right. \qquad (6.29)$$

$$\left. + \frac{1}{2}\sigma^2(K)K^2\nabla_K^2 P^D(S, \nu, K)\right]d\nu = \int_0^\infty pe^{-p\nu}\nabla_\nu P^D(S, \nu, K)d\nu$$

$$= -pP^D(S, 0, K) + p\int_0^\infty P^D(S, \nu, K)pe^{-p\nu}d\nu$$

$$= p\left[P(S, T, K) - P^D(S, 0, K)\right] = p\left[P(S, T, K) - P(S, 0, K)\right],$$

where in the last line we took into account Eq.(6.23).

Thus, finally $P(S, T, K)$ solves the following problem

$$-qP(S,T,K) - (r-q)K\nabla_K P(S,T,K) + \frac{1}{2}\sigma^2(K)K^2\nabla_K^2 P(S,T,K)$$
$$= \frac{P(S,T,K) - P(S,0,K)}{X(T)},$$
$$P(S,0,K) = (K-S)^+. \tag{6.30}$$

In contrast to the Dupire equation which belongs to the class of PDE, Eq.(6.30) is an ODE, or more precisely a partial divided-difference equation (PDDE), since the derivative in time in the right hands part is now replaced by a divided difference. In the form of an ODE it reads

$$\left[\frac{1}{2}\sigma^2(K)K^2\nabla_K^2 - (r-q)K\nabla_K - \left(q + \frac{1}{X(T)}\right)\right]$$
$$\times P(S,T,K) = -\frac{P(S,0,K)}{X(T)}. \tag{6.31}$$

This equation could be solved analytically for some particular form of the local volatility function $\sigma(K)$ which are considered in the next Section. Also in the same way a similar equation could be derived for the Call option price $C_0(S, T, K)$ which reads

$$\left[\frac{1}{2}\sigma^2(K)K^2\nabla_K^2 - (r-q)K\nabla_K - \left(q + \frac{1}{X(T)}\right)\right]$$
$$\times C_0(S,T,K) = -\frac{C_0(S,0,K)}{X(T)},$$
$$C_0(S,0,K) = (S-K)^+. \tag{6.32}$$

Solving Eq.(6.31) or Eq.(6.32) provides the way to determine $\sigma(K)$ given market quotes of Call and Put options with maturity T. However, this allows calibration of just a single term. Calibration of the entire local volatility surface, in principle, could be done term-by-term (because of the time-homogeneity assumption) if Eq.(6.31), Eq.(6.32) could be generalized to this case.

6.2.3 *PDDE for multiple terms*

This generalization can be done in the same way as presented in [Carr and Itkin (2018a)], Section 4. Therefore, we refer the reader to that Section while here we provide just some useful comments.

To address calibration of multiple smiles we need to relax the assumption about time-homogeneity of the D_t process defined in Eq.(6.1). We assume that the local variance $\sigma(D_t)$ is no more time-homogeneous, but a piecewise constant function of time $\sigma(D_t, t)$.

Let $T_1, T_2, \ldots, T_M$ be the time points at which the variance rate $\sigma^2(D_t)$ jumps deterministically. In other words, at the interval $t \in [T_0, T_1)$, the variance rate is $\sigma_0^2(D_t)$, at $t \in [T_1, T_2)$ it is $\sigma_1^2(D_t)$, etc. This can be also represented as

$$\sigma^2(D_t, t) = \sum_{i=0}^{M} \sigma_i^2(D_t) w_i(t), \tag{6.33}$$

$$w_i(t) \equiv \mathbf{1}_{t-T_i} - \mathbf{1}_{t-T_{i+1}}, \quad i = 0, \ldots, M, \quad T_0 = 0, \ T_{M+1} = \infty.$$

where the function $\mathbf{1}_x$ is defined in Eq.(5.31).

Note, that

$$\sum_{i=0}^{M} w_i(t) = \mathbf{1}_t - \mathbf{1}_{t-\infty} = 1, \quad \forall t \geq 0.$$

Therefore, in case when all $\sigma_i^2(D_t)$ are equal, ie, independent on index i, Eq.(6.33) reduces to the case considered in the previous Sections.

This implies that the volatility $\sigma(D_t)$ jumps as a function of time at the calendar times $T_0, T_1, \ldots, T_M$, and not at the business times ν determined by the Gamma clock. Otherwise, the volatility function would have been changed at random (business) times which means it is stochastic. But this definitely lies out of scope of our model. Therefore, we need to change Eq.(6.33) to

$$\sigma^2(D_t, t) = \sum_{i=0}^{M} \sigma_i^2(D) \bar{w}_i(\mathbb{E}_{\mathbb{Q}}(t)), \tag{6.34}$$

$$\bar{w}_i(\mathbb{E}_{\mathbb{Q}}(t)) = \mathbf{1}_{X^{-1}(t-T_i)} - \mathbf{1}_{X^{-1}(t-T_{i+1})}, \quad i = 0, \ldots, M,$$

$$X^{-1}(t) = \frac{1}{q-r} \log\left[1 - (r-q)t\right]. \tag{6.35}$$

As per the last line, $X(t)$ exists $\forall t \geq 0$ if $q > r$, and $\forall t < 1/(r-q)$ if $r > q$. Hence, when using Eq.(6.6) we have

$$\sigma^2(D_t, t)\Big|_{t=\Gamma_{X(t)}} = \sum_{i=0}^{M} \sigma_i^2(D) \bar{w}_i(X(t)) = \sum_{i=0}^{M} \sigma_i^2(D) w_i(t). \tag{6.36}$$

Accordingly, if the calendar time t belongs to the interval $T_0 \leq t < T_1$, the infinitesimal generator $\mathcal{A}$ of the semigroup $\mathcal{T}_\nu^D$ is a function of $\sigma(D_t)$ (and not of $\sigma(D_\nu)$). As at $T_0 \leq t < T_1$ we assume $\sigma(D) = \sigma_0(D)$, i.e. is constant in time, it doesn't depend on ν. Thus, $\mathcal{A}$ (which for this interval of time we will denote as $\mathcal{A}_0$) is still time-homogeneous.

Similarly, one can see, that for $T_1 \leq t < T_2$ the infinitesimal generator $\mathcal{A}_1$ of the semigroup $\mathcal{T}_\nu^D$ is also time-homogeneous and depends on $\sigma_1(D)$, etc.

Further, similar to [Carr and Itkin (2018a)] it could be shown that the forward partial divided difference equation for the Put price $P(S, T_i, K)$, $i = 1, \ldots, M$ reads

$$\left[\frac{1}{2}\sigma^2(K)K^2\nabla_K^2 - (r-q)K\nabla_K - \left(q + \frac{1}{X(T_i) - X(T_{i-1})} \right) \right] P(S, T_i, K)$$
$$= -\frac{P(S, T_{i-1}, K)}{X(T_i) - X(T_{i-1})}. \tag{6.37}$$

Here the local variance function $\sigma^2(K) = \sigma_i^2(K)$ as it corresponds to the interval $(T_{i-1}, T_i]$ where the above ODE is solved.

Eq.(6.37) is a recurrent equation that can be solved for all $i = 1, \ldots, M$ sequentially starting with $i = 1$ subject to some boundary conditions.

6.2.4 *Boundary conditions*

In many financial models where dynamics of the stock price is represented by a geometric Brownian motion (perhaps with local or stochastic volatility), for instance, the celebrated Black-Scholes model, the boundary condition at $K \to \infty$ is set to be

$$P(S, T_i, K) \to \mathcal{D}_i K - \mathcal{Q}_i S, \quad K \to \infty,$$

where $\mathcal{D}_i = e^{-rT_i}$ is the discount factor, and $\mathcal{Q}_i = e^{-qT_i}$. Indeed, as it could be easily checked this condition is a valid solution of the Dupire forward equation Eq.(6.28), and also reflects the fact that at $K \to \infty$ the Put option price should be linear in K. However, this boundary condition doesn't solve Eq.(6.31), so it could not be used in our model.

Therefore, we propose to setup the boundary condition at $K \to \infty$ by still assuming it to be a linear function of K of the form

$$\lim_{K \to \infty} P(S, T, K) = A(T)K - B(T)S, \tag{6.38}$$

where $A(T), B(T)$ are some functions of maturity T to be determined, so the expression in Eq.(6.38) solves Eq.(6.31).

Obviously, $T_0 = 0$ implies $A(T_0) = B(T_0) = 1$. Then we can proceed recursively. For the next given maturity $T = T_1$ plugging in Eq.(6.38) into Eq.(6.37) we obtain at $K \to \infty$

$$-(r-q)KA(T_1)p_1 - (p_1q + 1)(A(T_1)K - B(T_1)S) = -P(S, T_0, K),$$
$$P(S, T_0, K) = A(T_0)K - B(T_0)S = K - S, \tag{6.39}$$
$$p_j = X(T_j) - X(T_{j-1}) > 0.$$

From these equations we obtain

$$B(T_1) = \frac{1}{p_1 q + 1}, \qquad A(T_1) = \frac{1}{p_1 r + 1}. \tag{6.40}$$

So in this case $A(T_1), B(T_1)$ are an analog of some kind of discrete compounding.

Proceeding recursively, we derive a general relationship

$$B(T_i) = \frac{B(T_{i-1})}{p_i q + 1} = \frac{1}{\prod_{k=1}^{i}(p_i q + 1)}, \tag{6.41}$$

$$A(T_i) = \frac{A(T_{i-1})}{p_i r + 1} = \frac{1}{\prod_{k=1}^{i}(p_i r + 1)}, \qquad i = 1, \ldots, M.$$

Therefore, in our model the natural boundary conditions for the Put option price are

$$\begin{cases} P(S, T_i, K) = 0, & K \to 0, \\ P(S, T_i, K) = A(T_i)K - B(T_i)S \approx A(T_i)K, & K \to \infty, \end{cases} \tag{6.42}$$

A similar equation can be obtained for the Call option prices, which reads

$$\left[\frac{1}{2}\sigma^2(K)K^2 \nabla_K^2 - (r-q)K\nabla_K - \left(q + \frac{1}{X(T_i) - X(T_{i-1})} \right) \right] C(S, T_i, K)$$
$$= -\frac{C(S, T_{i-1}, K)}{X(T_i) - X(T_{i-1})}, \tag{6.43}$$

subject to the boundary conditions

$$\begin{cases} C(S, T_i, K) = B(T_i)S, & K \to 0, \\ C(S, T_i, K) = 0, & K \to \infty. \end{cases} \tag{6.44}$$

6.3 Piecewise models of local variance/volatility

To calibrate the local volatility surface by solving Eq.(6.37) we need to make further assumptions about the shape of the local volatility surface. To recall, we assume this surface to be piecewise constant in time. In the strike space [Carr and Nadtochiy (2017)] considered it to be a piecewise constant, while in [Carr and Itkin (2018a)] a piecewise linear local variance in the strike space was considered. As shown in [Carr and Itkin (2018a)], in those cases Eq.(6.37) can be solved in closed form.

Here we want to extend a class of local volatility models that allow a closed form solution. To proceed, we start by doing a change of the dependent variable from $P(S, T_j, K)$ to

$$V(S, T_j, K) = P(S, T_j, K) - [A(T_j)K - B(T_j)S]^+, \tag{6.45}$$

where V is known as a *covered* Put. This definition of V allows re-writing Eq.(6.37) in a more elegant form

$$- v_j(x)x^2 V_{x,x}(x) + b_{1,j}x V_x(x) + b_{0,j}V(x) = c_j(x), \qquad (6.46)$$

$$b_{1,j} = p_j(r - q), \quad b_{0,j} = p_j q + 1, \quad c_j(x) = V(S, T_{j-1}, x),$$

$$v_j(x) = p_j \sigma^2(x)/2,$$

where $V(x) = V(S, T_j, x)$ and $x = K/S$ is the inverse moneyness.

Accordingly, based on the definition of $V(x)$ and Eq.(6.42), the boundary conditions to Eq.(6.46) become homogeneous

$$\begin{cases} V(x) = 0, & x \to 0, \\ V(x) = 0, & x \to \infty. \end{cases} \qquad (6.47)$$

In the next sections we consider several popular approximations of the local volatility surface in the strike space. Each approximation assumes some functional form of the local volatility curve in the strike space, which is a strip of the volatility surface given time to maturity T. Thus, parameters of these approximations change with time. Also further on for simplicity we assume that $r > q > 0$, but this assumption could be easily relaxed.

6.3.1 *Local variance piecewise linear in a log-strike space*

Suppose that for each maturity T_j, $j \in [1, M]$ the market quotes are provided for a set of strikes $K_i, i = 1, \ldots, n_j$ where these strikes are assumed to be sorted in the increasing order. Then the corresponding continuous piecewise linear local variance function $\sigma_j^2(\chi)$ at the interval $[\chi_i, \chi_{i+1}]$, $\chi = \log K_i/S$, reads

$$v_{j,i}(\chi) = v_{j,i}^0 + v_{j,i}^1 \chi. \qquad (6.48)$$

Here we use the super-index 0 to denote a level v^0, and the super-index 1 to denote a slope v^1. Subindex $i = 0$ in $v_{j,0}^0, v_{j,0}^1$ corresponds to the interval $(0, \chi_1]$. Since $v_j(\chi)$ is a continuous function in χ, we have

$$v_{j,i}^0 + v_{j,i}^1 \chi_{i+1} = v_{j,i+1}^0 + v_{j,i+1}^1 \chi_{i+1}, \qquad i = 0, \ldots, n_j - 1. \qquad (6.49)$$

This means that the first derivative of $v_j(\chi)$ experiences a jump at points χ_i, $i \in Z \cap [1, n_j]$. As we assumed that $v(\chi, T)$ is a piecewise constant function of time, $v_{j,i}^0, v_{j,i}^1$ do not depend on T at the intervals $[T_j, T_{j+1})$, $j \in [0, M - 1]$, and jump to the new values at the points T_j, $j \in Z \cap [1, M]$.

A simple analysis shows that under this assumption by making a change of variables $x \mapsto \chi$, Eq.(6.46) could be transformed to

$$-v(\chi)V_{\chi,\chi}(\chi) + (b_1 + v(\chi))V_\chi(\chi) + b_0 V(\chi) = c(\chi), \qquad (6.50)$$

where for simplicity of notation we dropped index j.

This equation has the same type as that considered in [Itkin and Lipton (2018)], Section 2, and its solution could also be expressed in terms of confluent Hypergeometric functions, see [Polyanin and Zaitsev (2003)]

$$V(\chi) = C_1 y_1(\chi) + C_2 y_2(\chi) + I_{12}(\chi) \tag{6.51}$$

$$I_{12}(\chi) = y_2(\chi) \int \frac{y_1(\chi) c(\chi)}{(b_2 + a_2 \chi) W} d\chi - y_1(\chi) \int \frac{y_2 c(\chi)}{(b_2 + a_2 \chi) W} d\chi,$$

where $W = y_1(y_2)_\chi - y_2(y_1)_\chi$ is the so-called Wronskian of the fundamental solutions y_1, y_2. Thus, the problem is reduced to finding suitable fundamental solutions of the homogeneous version of Eq.(6.51). Based on [Polyanin and Zaitsev (2003)], if $a_2 \neq 0$ and $a_0 \neq 0$, the general solution reads

$$V(\chi) = (a_2 z)^{\beta_1 - 1} \mathcal{J}(\alpha_1, \beta_1, z), \tag{6.52}$$

$$z = \chi + \frac{b_2}{a_2}, \quad \alpha_1 = 1 + \frac{b_0 + b_1}{a_2}, \quad \beta_1 = 2 + \frac{b_1}{a_2}.$$

Here $\mathcal{J}(a, b, z)$ is an arbitrary solution of the degenerate Hypergeometric equation, i.e., Kummer's function, [Abramowitz and Stegun (1964)]. Two types of Kummer's functions are known, namely $M(a, b, z)$ and $U(a, b, z)$, which are Kummer's functions of the first and second kind.[4]

Accordingly, the approach of [Itkin and Lipton (2018)] can be directly applied to obtain a closed form solution of Eq.(6.51). In particular, in the vicinity of the origin the numerically satisfactory pair is, [Olver (1997)]

$$y_1(\chi) = (a_2 z)^{\beta_1 - 1} M(\alpha_1, \beta_1, z), \tag{6.53}$$

$$y_2(\chi) = (a_2)^{\beta_1 - 1} M(\alpha_1 - \beta_1 + 1, 2 - \beta_1, z).$$

$$W = a_2^{2\beta_1 - 2} e^z z^{\beta_1 - 2} \sin(\pi \beta_1) / \pi.$$

However, in the vicinity of infinity the numerically satisfactory pair is, [Olver (1997)]

$$y_1(\chi) = (a_2 z)^{\beta - 1} U(\alpha_1, \beta_1, z), \tag{6.54}$$

$$y_2(\chi) = e^z (a_2 z)^{\beta - 1} U(\beta_1 - \alpha_1, \beta_1, -z).$$

$$W = (-1)^{\alpha_1 - \beta_1} a_2^{2\beta_1 - 2} e^z z^{\beta_1 - 2}.$$

6.3.2 *Local variance piecewise linear in the strike space*

Another tractable model is where the local variance is piecewise linear in the strike space. In particular, this is the model we used in [Carr and Itkin (2018a)].

---

[4]Due to the linearity of the degenerate Hypergeometric equation any linear combination of Kummer's functions also solves this equation.

Similar to the previous section, the corresponding continuous piecewise linear local variance function $v_j(x)$ at the interval $[x_i, x_{i+1}]$ reads

$$v_{j,i}(x) = v_{j,i}^0 + v_{j,i}^1 x, \qquad (6.55)$$

where, however, it is now a function of x rather than χ. Since $v_j(x)$ is a continuous function in x, we have

$$v_{j,i}^0 + v_{j,i}^1 x_{i+1} = v_{j,i+1}^0 + v_{j,i+1}^1 x_{i+1}, \qquad i = 0, \dots, n_j - 1. \qquad (6.56)$$

This means that the first derivative of $v_j(x)$ experiences a jump at points x_i, $i \in [1, n_j]$. As we assumed that $v(x, T)$ is a piecewise constant function of time, $v_{j,i}^0, v_{j,i}^1$ don't depend on T at the intervals $[T_j, T_{j+1})$, $j \in 0, M-1]$, and jump to the new values at the points T_j, $j \in [1, M]$.

The Eq.(6.46) can be solved by induction. One starts with $T_0 = 0$, and at each time interval $[T_{j-1}, T_j]$, $j \in [1, M]$ solves the problem Eq.(6.46) for $V(x)$, and then obtains $P(S, T_j, x)$ from Eq.(6.45). Accordingly, the solution of Eq.(6.46) can be constructed separately for each interval $[x_{i-1}, x_i]$.

Substituting the representation Eq.(6.55) into Eq.(6.46), for the i-th spatial interval we obtain

$$-(b_2 + a_2 x)x^2 V_{x,x}(x) + b_1 x V_x(x) + b_{0,j} V(x) = c(x), \qquad (6.57)$$
$$b_2 = v_{j,i}^0, \quad a_2 = v_{j,i}^1.$$

Again, Eq.(6.57) is an *inhomogeneous* ordinary differential equation, and its solution can be represented in the form of Eq.(6.51) with

$$I_{12}(x) = -y_2(x) \int \frac{y_1(x)c(x)}{(b_2 + a_2 x)x^2 W(x)} dx + y_1(x) \int \frac{y_2(x)c(x)}{(b_2 + a_2 x)x^2 W(x)} dx$$
$$\equiv J_1 + J_2. \qquad (6.58)$$

The corresponding homogeneous equation can be solved as follows. First, if $b_2 \neq 0$ we make a change of independent variable $x \mapsto z = -a_2 x / b_2$. As the result the homogeneous Eq.(6.57) takes the form

$$b_2(z-1)z V_{z,z}(z) + b_1 z V_z(z) + b_0 V(z) = 0. \qquad (6.59)$$

Then we make a change of the dependent variable $V(z) \mapsto z^m G(z)$ with m being some constant for the given time slice. This leads to the equation

$$z^m[\gamma + b_2(m-1)mz]G(z) + z^{m+1}[b_1 + 2b_2 m(z-1)]G'(z) \qquad (6.60)$$
$$+ b_2(z-1)z^{m+2}G''(z) = 0,$$
$$\gamma = b_0 + m(b_2 + b_1 - b_2 m).$$

Next we solve for m which makes γ vanish, to obtain

$$m^{\pm} = \frac{b_2 + b_1 \pm \sqrt{4b_2b_0 + (b_2 + b_1)^2}}{2b_2}. \qquad (6.61)$$

It is worth mentioning that if the determinant D in this expression is negative, both m^+, m^- become complex. However, this is not a problem for the solution as coefficients C_1, C_2 in Eq.(6.51) could be complex as well, and such that the Put price is real.

Substituting this into Eq.(6.60) and rearranging we obtain

$$-m(m-1)G(z) + \left(2m - \frac{b_1}{b_2} - 2mz\right) G'(z) + z(1-z)G''(z) = 0, \qquad (6.62)$$
$$m \in [m^+, m^-],$$

which is a Hypergeometric equation. As m can take two values, we need to choose the right one such that the final solution would obey the boundary conditions.

Combining all the above steps, the solution of Eq.(6.59) could be written as

$$y_1(x) = z^m \left[{}_2F_1\left(m-1, m, c; z\right) \right], \qquad (6.63)$$
$$y_2(x) = z^m \left[z^{1-c}\, {}_2F_1\left(m-c, m+1-c, 2-c; z\right) \right],$$
$$m = m^+, \quad c = 2m - \frac{b_1}{b_2}, \quad z = -\frac{a_2}{b_2}x,$$

Here ${}_2F_1(a, b, c; z)$ is the ordinary Hypergeometric function, [Olver (1997)]. It has regular singularities at $z = 0, 1, \infty$. In terms of the solution in Eq.(6.52), these singularities correspond to $K = 0$, $v = 0$ and $K \to \infty$. We will show below that at $K \to \infty$ the coefficient a_2 for this interval is usually positive, so the variance is positive. However, the sign of b_2 could be both plus and minus. Therefore, if $b_2 > 0$ at this interval, we have $x \to \infty$, $z \to -\infty$. If $b_2 < 0$ at this interval, we have $x \to \infty$, $z \to \infty$.

When none of $c, c-a-b, a-b$ is an integer, we have a pair of fundamental solutions $f_1(x), f_2(x)$ that in Eq.(6.52) are represented by expressions in square brackets. It is known that this pair is numerically satisfactory, [Olver (1997)] aside of singularities at $z = 1$ and $z \to \infty$. Wronskian of these fundamental solutions $W(f_1(x), f_2(x))$ is

$$W(f_1(x), f_2(x)) = (1-c)z^{-c}(1-z)^{c-2m}, \quad z = -a_2x/b_2.$$

Accordingly,

$$W(y_1(x), y_2(x)) = -\frac{a_2(1-c)}{b_2} z^{2m-c}(1-z)^{c-2m}, \quad z = -a_2x/b_2. \qquad (6.64)$$

In the vicinity of **singularity at** $z = 1$ this pair, however, is not numerically satisfactory. Then we have to use another solution of Eq.(6.62) which is, [Olver (1997)]

$$y_1(x) = z^m \left[{}_2F_1 \left(m - 1, m, 2m - c; 1 - z \right) \right], \tag{6.65}$$

$$y_2(x) = z^m \left[(1 - z)^{c-2m+1} {}_2F_1 \left(c - m + 1, c - m, c - 2m + 2; 1 - z \right) \right],$$

$$W(y_1(x), y_2(x)) = -\frac{a_2(2m - 1 - c)}{b_2} (1 - z)^{c-2m} z^{2m-c}, \quad z = -a_2 x / b_2.$$

The numerically satisfactory fundamental solutions in the vicinity of **singularity at** $z = \infty$ are described below.

6.3.2.1 *Numerically satisfactory solutions of Eq.(6.59) at $z \to \infty$*

According to [Olver (1997)], the numerically satisfactory fundamental solutions of Eq.(6.59) in the vicinity of **singularity at** $z = \infty$ are

$$y_1(x) = z^m [z^{-A} {}_2F_1 \left(A, A - C + 1, A - B + 1, 1/z \right)], \tag{6.66}$$

$$y_2(x) = z^m [z^{-B} {}_2F_1 \left(B, B - C + 1, B - A + 1; 1/z \right)],$$

where in our case $A = m - 1, B = m, C = c$. This substitution transforms the second solution in Eq.(6.66) to

$$y_2(x) = z^m [z^{-m} {}_2F_1 \left(m, m - c + 1, 2; 1/z \right)], \tag{6.67}$$

and behaves well at $z \to \infty$. However, since in our setting $n \equiv A - B + 1 = m - 1 - m + 1 = 0$, and due to the property

$$\lim_{c \to -n} \frac{F(a, b, c; z)}{\Gamma(c)} = \frac{(a)_{n+1}(b)_{n+1}}{(n+1)!} z^{n+1} F(a + n + 1, b + n + 1, n + 2; z),$$

$$y_1(x) = F(m - 1, m - c, 0; z)$$

$$= \Gamma(0) \frac{(m - 1)_1 (m - c)_1}{(1)!} z F(m, m - c + 1, 2; z),$$

it turns out that the first solution differs from the second one just by a constant multiplier, i.e. they are not independent. Therefore, in this case instead the first solution $y_1(x)$ should be chosen based on a more sophisticated analytic continuation of the Hypergeometric function, [Bateman and Erdélyi (1953)].

$$y_1(x) = z^m [(-z)^{1-m} \frac{\Gamma(c)}{\Gamma(m)\Gamma(c - m + 1)} \Psi(z)], \quad |z| > 1, \ |\mathrm{ph}(-z)| < \pi,$$

$$\Psi(z) = 1 - \frac{1}{z} \sum_{k=0}^{\infty} \frac{(m - 1)_{k+1}(m - c)_{k+1}}{k!(k+1)!} z^{-k} \left[\log(-z) + \phi_k \right], \tag{6.68}$$

$$\phi_k \equiv \psi(k+1) + \psi(k+2) - \psi(m+k) - \psi(c-m-k),$$
$$(m)_k = \Gamma(m)/\Gamma(k), \quad \psi(x) = \Gamma'(x)/\Gamma(x).$$

■

However, we cannot use this solution at $z \to \infty$ as well as to use the solution in Eq.(6.63) at $z \to 0$. This is caused by the Roger Lee's moment matching formula, [Lee (2004)] which states that in the wings the implied variance surface should be at most linear in the normalized strike (or log-strike). It is also shown in [De Marco *et al.* (2013); Gerhold and Friz (2015)], that the asymptotic behavior of the local variance is linear in the log strike at both $K \to \infty$ and $K \to 0$. While the result for $K \to 0$ is shown to be true at least for the Heston and Stein-Stein models, the result for $K \to \infty$ directly follows from Lee's moment formula for the implied variance v_{BS} and the representation of σ^2 via the total implied variance $w = v_{BS}T$ in Eq.(1.30).

Thus, the considered model of the local variance linear in strike is not applicable at the first $0 \le x \le x_1$ and the last $x_{n_j} < x < \infty$ strike intervals for every smile $T = T_j$ as it violates Lee's formula. Therefore, at these two intervals we use the model discussed in Section 6.3.1 where the local variance is linear in the log-strike.

It is interesting to mention, that in [Itkin and Lipton (2018); Carr and Itkin (2018a)] and in section 6.3.1 the closed form solution was obtained in terms of Kummer's functions. Here the solution is expressed via Hypergeometric functions $_2F_1(a, b, c; x)$.

As two solutions $y_1(x)$ and $y_2(x)$ are independent, Eq.(6.51) is a general solution of Eq.(6.57). Two constants C_1, C_2 should be determined based on the boundary conditions for the function $y(x)$.

The boundary conditions for the ODE Eq.(6.57) in the x space at zero and infinity are given in Eq.(6.47), i.e. they are homogeneous. Based on the usual shape of the local variance curve and its positivity, for $x \to 0$, we expect that $v_{j,i}^1 < 0$. Similarly, for $x \to \infty$ we expect that $v_{j,i}^1 > 0$. In between these two limits the local variance curve for a given maturity T_j is assumed to be continuous, but the slope of the curve could be both positive and negative, see, e.g., [Itkin (2015)] and references therein.

6.3.3 *Local volatility piecewise linear in the strike space*

Another popular model is where the local volatility is assumed to be piecewise linear in the strike space. This model previously was frequently considered in the literature, e.g., [Hull and White (2015); Kienitz and Caspers (2017)]. Below we show that with this assumption our model remains tractable, and a closed form solution can be obtained by using the same approach as elaborated on in [Itkin and Lipton (2018); Carr and Itkin (2018a)].

Accordingly, the corresponding continuous piecewise linear local volatility function $\sigma_j(x)$ on the interval $[x_i, x_{i+1}]$ reads

$$\sigma_{j,i}(x) = \sigma_{j,i}^0 + \sigma_{j,i}^1 x, \qquad (6.69)$$

Since $\sigma_j(x)$ is a continuous function in x, we have

$$\sigma_{j,i}^0 + \sigma_{j,i}^1 x_{i+1} = \sigma_{j,i+1}^0 + \sigma_{j,i+1}^1 x_{i+1}, \qquad i = 0, \ldots, n_j - 1. \qquad (6.70)$$

Again, this means that the first derivative of $\sigma_j(x)$ experiences a jump at points x_i, $i \in [1, n_j]$. As $\sigma(x,T)$ is a piecewise constant function of time, $\sigma_{j,i}^0, \sigma_{j,i}^1$ do not depend on T at the intervals $[T_j, T_{j+1})$, $j \in [0, M-1]$, and jump to the new values at the points T_j, $j \in [1, M]$.

Substituting the representation Eq.(6.69) into Eq.(6.46), for the i-th spatial interval we obtain

$$-(b_2 + a_2 x)^2 x^2 V_{x,x}(x) + b_1 x V_x(x) + b_{0,j} V(x) = c(x), \qquad (6.71)$$
$$b_2 = \sigma_{j,i}^0, \quad a_2 = \sigma_{j,i}^1.$$

Again, Eq.(6.71) is an *inhomogeneous* ordinary differential equation, and its solution can be represented in the form of Eq.(6.51) with

$$I_{12}(x) = -y_2(x) \int \frac{y_1(x)c(x)}{(b_2 + a_2 x)^2 x^2 W(x)} dx + y_1(x) \int \frac{y_2(x)c(x)}{(b_2 + a_2 x)^2 x^2 W(x)} dx$$
$$\equiv L_1 + L_2.$$

The corresponding homogeneous equation can be solved as follows. First, if $b_2 \neq 0, b_2 + a_2 x \neq 0$ we make a change of independent variable $x \mapsto z = a_2 b_1 x / [b_2^2 (b_2 + a_2 x)]$. As the result the homogeneous Eq.(6.71) takes the form

$$b_2 z^2 (-b_1 + b_2^2 z) V_{z,z}(z) + z \left[2 b_2^4 z + (b_1 - b_2^2 z)^2 \right] V_z(z) + b_0 (b_1 - b_2^2 z) V(z) = 0.$$

Next we make a change of the dependent variable

$$V(z) \mapsto z^{k_1} \left(\frac{z}{b_2^2 z + b_1} \right)^{k_2} G(z)$$

with k_1, k_2 being some constants for the given time slice. This leads to the equation

$$0 = -b_2^2 z \left(b_1 - b_2^2 z\right)^2 G''(z) + f_1(z)G'(z) + f_0(z)G(z), \qquad (6.72)$$
$$f_1(z) = z \left(b_1 - b_2^2 z\right) \left[b_2^4 z(2k_1 + z + 2) - 2b_2^2 b_1(k_1 + k_2 + z) + b_1^2\right],$$
$$f_0(z) = q_0 + q_1 z + q_2 z^2 - b_2^6 k_1 z^3,$$
$$q_2 = b_2^4 \left[b_0 - b_2^2 k_1(k_1 + 1) + b_1(3k_1 + k_2)\right],$$
$$q_1 = b_2^2 b_1 \left[2b_2^2 k_1(k_1 + k_2) - 2b_0 - b_1(3k_1 + 2k_2)\right],$$
$$q_0 = b_1^2 \left[b_0 - (k_1 + k_2)\left(b_2^2(k_1 + k_2 - 1) - b_1\right)\right].$$

We now request that $f_0(z)$ is proportional to $z\left(b_1 - b_2^2 z\right)^2$ with some constant multiplier q, i.e.

$$f_0(z) = qz \left(b_1 - b_2^2 z\right)^2.$$

Solving this equation term by term in powers of z, we obtain

$$k_1 = -\frac{q}{b_2^2}, \quad k_2 = \frac{q(b_1 + q) - b_2^2(b_0 + q)}{b_2^2 b_1},$$
$$q = \frac{1}{2}\left(b_2^2 - b_1 \pm \sqrt{b_2^4 + 2b_2^2(2b_0 + b_1) + b_1^2}\right).$$

Accordingly, substituting these definitions into Eq.(6.72) one finds

$$0 = zG''(z) + (b + z)G'(z) - aG(z),$$
$$b = 2 - \frac{b_1 + 2q}{b_2^2}, \quad a = \frac{q}{b_2^2}.$$

This is a sort of Kummer equation which has two independent solutions, [Polyanin and Zaitsev (2003)]

$$G(z) = e^{-z}U(a + b, b, z), \quad G(z) = e^{-z}M(a + b, b, z). \qquad (6.73)$$

Accordingly, as q can take two values corresponding to the plus and minus sign, we have four fundamental solutions of the original equation Eq.(6.72).

Similar to the previous section, we cannot use these solutions at the first $0 \le x \le x_1$ and the last $x_{n_j} < x < \infty$ strike intervals for every smile $T = T_j$ as it violates Lee's formula. Therefore, at these two intervals we use the model discussed in Section 6.3.1 where the local variance is linear in the log-strike. Accordingly, the local volatility is a square root of the local variance.

6.4 Computation of the source term

Computation of the source term pI_{12} in Eq.(6.51) could be achieved in several ways. The most straightforward one is to use numerical integration since the Put price $P(x, T_{i-1})$ as a function of x is already known when we solve Eq.(6.51) for $T = T_i$. We underline that this is not the case in [Itkin and Lipton (2018)], because there the function $P(x, T_{i-1})$ is obtained by using an inverse Laplace transform, and as such is known only for a discrete set of strikes at the previous time level. Therefore, some kind of interpolation is necessary to find the local variance at all strikes when doing integration. Moreover, this interpolation must preserve no-arbitrage, see [Itkin and Lipton (2018)].

On the other hand, using no-arbitrage interpolation provides another advantage, as it makes it possible to compute the source term integrals in closed form if the interpolating function is wisely chosen. Here we want to exploit the same idea, thus significantly improving computational performance of our model as compared with the numerical integration.

Below as an example consider the case of the local variance piecewise linear in the strike space. Then based on solutions found in Section 6.3.2 in Eq.(6.63) we have

$$J_1(x) = -y_2(x) \int \frac{y_1(x)c(x)}{(b_2 + a_2 x)x^2 W(x)} dx = -y_2(x)\frac{a_2^2}{b_2^3} \int \frac{y_1(z)c(z)}{(1-z)z^2 W(z)} dz,$$

$$(6.74)$$

$$y_1(z) = z^m {}_2F_1 (m - 1, m, c; z), \quad c(z) = V(S, T_{j-1}, z), \quad z = -a_2 x/b_2,$$

where $W(z)$ is defined in Eq.(6.64).

Following the idea of [Itkin and Lipton (2018)], in [Carr and Itkin (2018a)]) we introduced a non-linear interpolation

$$P(x) = \gamma_0 + \gamma_2 x^2, \quad x_1 \le x \le x_3, \qquad (6.75)$$

$$\gamma_0 = \frac{P(x_3)x_1^2 - P(x_1)x_3^2}{x_1^2 - x_3^2}, \qquad \gamma_2 = \frac{P(x_1) - P(x_3)}{x_1^2 - x_3^2}.$$

Then Proposition 6.1 in [Carr and Itkin (2018a)] proves that this interpolation scheme is arbitrage-free.

It is worth emphasizing that the proposed interpolation doesn't affect the solution values (quotes) at given market strikes since the piecewise interpolator is constructed to exactly match those values. So the interpolation only affects the Put values that are not known, i.e., those with strikes that lie in between the given market strikes. Therefore, if these strikes are not

used, i.e. in trading or hedging, the influence of the interpolation is unob-
servable at all. If, however, they are used for some purpose, the difference
with the exact solution is small (within the error of interpolation), while
the approximate solution for these strikes yet preserves no-arbitrage.

Recall, that we introduced $V(x)$ using Eq.(6.45). Accordingly, computation of the term $c(z)$ in Eq.(6.74) is given below.

6.4.1 *No-arbitrage interpolation at $z \to 1$*

As by definition in Eq.(6.63) $z = -\frac{a_2}{b_2}x$, this implies that

$$1 - z = 1 + \frac{a_2}{b_2}x = \frac{v_{ji}}{b_2}.$$

Obviously, $v_{ji} \geq 0$. Therefore, when z is close to 1, two situations are
possible:

(1) $z < 1$, which implies $b_2 > 0$, and accordingly $a_2 < 0$;
(2) $z > 1$, which implies $b_2 < 0$, and accordingly $a_2 > 0$.

Suppose for interpolation of the Put price we use Eq.(6.75), i.e.

$$P(x) = \gamma_0 + \gamma_2 x^2, \quad x_1 \leq x \leq x_3, \tag{6.76}$$

$$\gamma_0 = \frac{P(x_3)x_1^2 - P(x_1)x_3^2}{x_1^2 - x_3^2} = P_1 - \frac{P_3 - P_1}{x_3^2 - x_1^2}x_1^2 > 0,$$

$$\gamma_2 = \frac{P(x_1) - P(x_3)}{x_1^2 - x_3^2} > 0.$$

The second inequality is obvious since $P(x_3) > P(x_1)$ if $x_3 > x_1$. The first
one follows from the fact that the Put price exceeds its intrinsic value, i.e.

$$P_i = [A(T_j)K_i - B(T_j)S]^+ + \varepsilon_i, \qquad \varepsilon_i > 0.$$

Suppose, e.g., that both strikes K_1, K_3 are in-the-money. Then

$$\gamma_0 = P_1 - \frac{P_3 - P_1}{x_3^2 - x_1^2}x_1^2 = P_1 - \frac{A(T_j)S(x_3 - x_1) + \varepsilon_3 - \varepsilon_1}{x_3^2 - x_1^2}x_1^2 \tag{6.77}$$

$$= \frac{P_1 x_3 + x_1(P_1 - A(T_j)K_1)}{x_3 + x_1} + \frac{\varepsilon_1 - \varepsilon_3}{x_3^2 - x_1^2}x_1^2 > 0,$$

as based on the properties of the Put price $\varepsilon_1 > \varepsilon_3$.

From Eq.(6.76) it follows that

$$V = \gamma_0 + \gamma_2 x^2 - A(T_j)Sx + B(T_j)S = \bar{\gamma}_0 + \gamma_1 z + \bar{\gamma}_2 z^2, \tag{6.78}$$

$$\bar{\gamma}_0 = \gamma_0 + +B(T_j)S, \quad \gamma_1 = \frac{a_2}{b_2}A(T_j)S, \quad \bar{\gamma}_2 = \gamma_2 \frac{a_2^2}{b_2^2}.$$

It was proven in [Carr and Itkin (2018a)] that interpolation Eq.(6.76) preserves no-arbitrage, and so that in Eq.(6.78). We use it when computing $\mathcal{J}_2(x)$ in Eq.(6.99). ∎

It turns out that now the integral in Eq.(6.74) can be computed in closed form. Indeed

$$\int \frac{y_1(z)c(z)}{(1-z)z^2W(z)}dz = I_0 + I_1 + I_2, \qquad (6.79)$$

$$I_0 = \gamma_0 \int \frac{y_1(z)}{(1-z)z^2W(z)}dz$$
$$= \bar{\gamma}_0 A(z)\frac{1}{\Gamma(c)(c-m-1)}{}_2F_1\left(c-m-1, c-m+1, c, z\right),$$

$$I_1 = \gamma_1 \int \frac{zy_1(z)}{(1-z)z^2W(z)}dz$$
$$= \gamma_1 z A(z)\frac{1}{\Gamma(c)(c-m)}{}_2F_1\left(c-m, c-m, c, z\right)$$

$$I_2 = \bar{\gamma}_2 \int \frac{z^2 y_1(z)}{(1-z)z^2W(z)}dz$$
$$= \bar{\gamma}_2 A(z)z^2\frac{1}{(c-m+1)\Gamma(c)}{}_3F_2\left[\begin{matrix}c-m, & c-m+1, & c-m+1 \\ & c, & 2+c-m\end{matrix}; z\right],$$

$$A(z) = \frac{b_2}{a_2}\Gamma(c-1)z^{c-m-1},$$

where ${}_3F_2\left[\begin{matrix}a_1, a_2, a_3 \\ b_1, b_2\end{matrix}; z\right]$ is a generalized Hypergeometric function ([Askey and Daalhuis (2010)]).

The second integral in the definition of J_2

$$J_2(x) = y_1(x)\int \frac{y_2(x)c(x)}{(b_2+a_2x)x^2W(x)}dx = y_1(x)\frac{a_2^2}{b_2^3}\int \frac{y_2(z)c(z)}{(1-z)z^2W(z)}dz, \qquad (6.80)$$

$$y_2(z) = z^{m+1-c}{}_2F_1\left(m-c, m+1-c, 2-c; z\right),$$

could be computed in a similar way. The result reads

$$\int \frac{y_2(z)c(z)}{(1-z)z^2W(z)}dz = \mathcal{I}_0 + \mathcal{I}_1 + \mathcal{I}_2, \qquad (6.81)$$

$$\mathcal{I}_0 = \gamma_0 \int \frac{y_2(z)}{(1-z)z^2W(z)}dz = \bar{\gamma}_0 A(z)\frac{1}{m}{}_2F_1\left(2-m, -m, 2-c, z\right),$$

$$\mathcal{I}_1 = \gamma_1 \int \frac{zy_2(z)}{(1-z)z^2W(z)}dz$$

$$= \gamma_1 A(z) z \frac{1}{(m-1)} {}_2F_1 \left(1-m, 1-m, 2-c, z\right),$$

$$\mathcal{I}_2 = \bar{\gamma}_2 \int \frac{z^2 y_2(z)}{(1-z)z^2 W(z)} dz$$

$$= \bar{\gamma}_2 A(z) z^2 \frac{1}{(m-2)} {}_3F_2 \left[\begin{matrix} 1-m, \ 2-m, \ 2-m \\ 2-c, \ 3-m \end{matrix} ; z \right],$$

$$A(z) = \frac{b_2}{a_2} \frac{\Gamma(1-c)}{\Gamma(2-c)} z^{-m},$$

Two special cases are the first $0 \le x \le x_1$ and the last $x_{n_j} < x < \infty$ intervals where the solution is given by Eq.(6.53) and Eq.(6.54).

6.4.2 Last interval $x_{n_j} \le x < \infty$

Since the right edge of this interval lies at infinity, the interpolation scheme in Eq.(6.75) should be slightly modified. This could be done twofold. The first option is to move the boundary from infinity to any very large but finite positive strike. Then the scheme in Eq.(6.75) could be used with no problem. But in our case it turns out that we are not able to compute these integrals in closed form. Therefore, we use another approach which consists in replacing the quadratic form in Eq.(6.75) with another nonlinear interpolation

$$c(\chi) = V(\chi, T_{j-1}, S) = \gamma_\infty z^{-\nu}, \quad z = \chi + \frac{b_2}{a_2}, \tag{6.82}$$

where $\gamma_\infty > 0$, $\nu > 0$ are some constants to be determined. Obviously, at $\chi \to \infty$ this interpolation preserves the correct boundary value of V as in Eq.(6.47), i.e. $V(\chi)$ vanishes in this limit. Derivation of the appropriate values of γ_∞, ν and a proof that the proposed interpolation preserves no-arbitrage are given in the next Section.

6.4.2.1 No-arbitrage interpolation at $\chi \to \infty$

In this Appendix we prove the following Proposition:

Proposition 6.1. *Recall that according to Eq.(6.82) the proposed interpolation scheme for $V(\chi, T_{j-1}, S)$ at the interval $x_{n_j} \le x < \infty$ reads*

$$c(\chi) = V(\chi, T_{j-1}, S) = \gamma_\infty z^{-\nu}, \quad z = \chi + \frac{b_2}{a_2}, \tag{6.83}$$

where $\gamma_\infty > 0$, $\nu > 0$ are some constants determined below in the proof. Also this scheme preserves no-arbitrage.

Proof. By construction, at $K \to \infty$, $c(\chi)$ converges to the correct boundary condition, i.e. vanishes. Assuming that K_{n_j} is in-the-money, Eq.(6.82) can be re-written in the form

$$P(K) = A(T_{j-1})K - B(T_{j-1})S + \gamma_\infty[\log(K/S) + b_2/a_2]^{-\nu}. \quad (6.84)$$

As at this interval $v = b_2 + a_2 \log(K/S) > 0$, and it was assumed that $K > S$, we must have $a_2 > 0$. Accordingly, to have a positive Put price we require $\gamma_\infty > 0$. This constant could be determined by using a known Put value at K_{n_j}, i.e. $P(K_{n_j}) = P_{n_j}$. This yields

$$\gamma_\infty = [P_{n_j} - A(T_{j-1})K_{n_j} - B(T_{j-1})S]\left[\frac{b_2}{a_2} + \log\left(\frac{K_{n_j}}{S}\right)\right]^\nu > 0. \quad (6.85)$$

Therefore, this definition is also consistent with the requirement of positiveness of γ_∞.

As this is described in detail in [Itkin and Lipton (2018)], the no-arbitrage conditions for the Put price read

$$P > 0, \quad P_K > 0, \quad P_{K,K} > 0.$$

Differentiating Eq.(6.84) on K, and then again, we obtain

$$P'_K = A(T_{j-1}) - \frac{\gamma_\infty \nu}{K}\left[\frac{b_2}{a_2} + \log\left(\frac{K}{S}\right)\right]^{-1-\nu}, \quad (6.86)$$

$$P''_K = \frac{\gamma_\infty \nu}{a_2 K^2}\left[\frac{b_2}{a_2} + \log\left(\frac{K}{S}\right)\right]^{-\nu-2}[b_2 + a_2(1 + \nu + \log(K/S))].$$

Analyzing these expressions we conclude that $P''_K > 0$. Observe that at $K \to \infty$ we also have $P'_K > 0$. Also observe that P'_K is a monotone function of K. Therefore, let us look at $P'_K(K_{n_j})$. Substitution of $K = K_{n_j}$ into the first line of Eq.(6.86) yields

$$P'_K(K_{n_j}) = A(T_{j-1}) + \frac{a_2\nu}{K_{n_j}(b_2 + a_2 \log(K/S)}\left[A(T_{j-1})K_{n_j} \quad (6.87)\right.$$
$$\left. - B(T_{j-1})S - P_{n_j}\right].$$

As the Put value exceeds its intrinsic value, $P'_K(K_{n_j})$ is positive if

$$0 < \nu < A(T_{j-1})K_{n_j}\left[\frac{b_2}{a_2} + \log\left(\frac{K_{n_j}}{S}\right)\right] \quad (6.88)$$
$$\times \left[P_{n_j} - A(T_{j-1})K_{n_j} + B(T_{j-1})S\right]^{-1} \equiv \Omega.$$

At large K_{n_j} the expression in the first square brackets is large, and in the second ones — small. Thus the upper boundary for ν is high enough.

Finally, we take into account the well-known upper bound of the Put option price which is, [Hull (1997)]

$$P_{n_j} \leq A(T_j)K_{n_j}.$$

Because of that, we can re-write Eq.(6.88) as

$$0 < \nu < \frac{A(T_{j-1})}{B(T_{j-1})} \frac{K_{n_j}}{S} \left[\frac{b_2}{a_2} + \log\left(\frac{K_{n_j}}{S}\right)\right] \approx \frac{K_{n_j}}{S} \left[\frac{b_2}{a_2} + \log\left(\frac{K_{n_j}}{S}\right)\right] \leq \Omega.$$
(6.89)

Therefore, if ν is chosen according to Eq.(6.88) or Eq.(6.89), this guarantees that $P'_K(K_{n_j}) > 0$. As $P'_K(K)$ is a monotone function of K, this proves that with this choice of ν the condition $P'_K(K) > 0$ is valid at the whole interval $x_{n_j} \leq x < \infty$. Thus, this interpolation preserves no-arbitrage. □

Recall that at this interval we assume the local variance to be linear in the log-strike χ. Therefore, the numerically stable pair of solutions of Eq.(6.51) is given in Eq.(6.54). Then the integral in Eq.(6.51) can be computed in closed form. In doing so we use the following notation from [Ng and Geller (1970)]

$$\int e^{-\alpha z} z^\nu U(a, b, z) dz = U_\nu(\alpha; a, b, z),$$

$$\int e^{-\alpha z} z^\nu M(a, b, z) dz = M_\nu(\alpha; a, b, z).$$

Then

$$I_{12}(\chi) = y_2(\chi) \int \frac{y_1(\chi)c(\chi)}{(b_2 + a_2\chi)W} d\chi - y_1(\chi) \int \frac{y_2(\chi)c(\chi)}{(b_2 + a_2\chi)W} d\chi,$$
(6.90)

$$\int \frac{y_1(\chi)c(\chi)}{(b_2 + a_2\chi)W} d\chi = \xi_\infty \int e^{-z} z^{-\nu} U(\alpha_1, \beta_1, z) dz = \xi_\infty U_{-\nu}(-1; \alpha_1, \beta_1, z),$$

$$\int \frac{y_2(\chi)c(\chi)}{(b_2 + a_2\chi)W} d\chi = \xi_\infty \int z^{-\nu} U(\beta_1 - \alpha_1, \beta_1, -z) dz$$

$$= (-1)^{-\nu} \xi_\infty U_{-\nu}(0; \beta_1 - \alpha_1, \beta_1, -z),$$

$$\xi_\infty = (-1)^{\beta_1 - \alpha_1} \gamma_\infty a_2^{2-\beta_1}.$$

As per [Ng and Geller (1970)],

$$M_\nu(-1; a, b, z) = e^{i\pi(\nu+1)} M_\nu(0; b - a, b, -z), \qquad (6.91)$$

$$M_\nu(0; a, b, z) = \frac{z^{\nu+1}}{\nu+1} {}_2F_2 \begin{bmatrix} \nu_1 + 1, \ a \\ \nu + 2, \ b \end{bmatrix}; z \Big],$$

$$b \neq 0, -1, -2, \ldots, \quad \nu \neq -1, -2, \ldots,$$

$$M_{-1}(0; a, b, z) = \frac{a}{b} z \ {}_3F_3 \begin{bmatrix} a + 1, \ 1, \ 1 \\ b + 1, \ 2, \ 3 \end{bmatrix}; z \Big] + \log(z),$$

$$U_\nu(\alpha; a, b, z) = \frac{\pi}{\sin(\pi b)} \Bigg[\frac{M_\nu(\alpha; a, b, z)}{\Gamma(1 + a - b)\Gamma(b)} - \frac{M_{\nu+1-b}(\alpha; 1 + a - b, 2 - b, z)}{\Gamma(a)\Gamma(2 - b)} \Bigg].$$

Therefore, all necessary integrals could be expressed in terms of generalized Hypergeometric functions. Alternatively, these integrals could be represented as

$$U_{-\nu}(-1; \alpha_1, \beta_1, z) = G_{2,3}^{2,1} \left(\begin{matrix} 1, \ 2+\alpha_1-\beta_1-\nu \\ 1-\nu, \ 2-\beta_1-\nu, \ 0 \end{matrix} \bigg| z \right), \qquad (6.92)$$

$$U_{-\nu}(0; \alpha_1, \beta_1, -z) = \frac{z^{1-\nu}}{\Gamma(1 - \alpha_1)\Gamma(\beta_1 - \alpha_1)} G_{2,3}^{2,2} \left(\begin{matrix} \nu, \ 1+\alpha_1-\beta_1 \\ 0, \ 1-\beta_1, \ \nu-1 \end{matrix} \bigg| -z \right),$$

where $G_{p,q}^{m,n} \left(\begin{matrix} a_1,\ldots,a_p \\ b_1,\ldots,b_q \end{matrix} \big| z \right)$ is the Meijer G-function, see [Olver (1997)].

It is not difficult to verify that at $K \to \infty$, and so $z \to \infty$, the integral $I_{12}(\chi)$ vanishes.

6.4.3   *First interval $0 \leq x \leq x_1$*

Recall that at this interval we assume the local variance to be linear in the log-strike χ. Since at $K \to 0$ we have $\chi \to -\infty$, the numerically stable pair of solutions of Eq.(6.51) is still given by Eq.(6.54).

However, at this interval we need another interpolation scheme because the previously described schemes don't give rise to tractable integrals. However, this could be achieved by using, e.g., the following nonlinear interpolation

$$c(\chi) = V(\chi, T_{j-1}, S) = \omega_0 e^z/z, \quad z = \chi + \frac{b_2}{a_2}, \qquad (6.93)$$

where $\omega_0 < 0$ is a constant to be determined. Obviously, at $K \to 0$, and so $z \to -\infty$, this interpolation preserves the correct boundary value of V as in Eq.(6.47), i.e. $V(\chi)$ vanishes in this limit. Derivation of the appropriate value of ω_0 and a proof that the proposed interpolation preserves no-arbitrage are given below.

6.4.3.1 *No-arbitrage interpolation at* $\chi \to -\infty$

In this Appendix we prove the following Proposition:

Proposition 6.2. *Recall that according to Eq.(6.82) the proposed interpolation scheme for* $V(\chi, T_{j-1}, S)$ *at the interval* $-\infty \leq x < x_1$ *reads*

$$V(\chi, T_{j-1}, S) = \omega_0 e^z / z, \quad z = \chi + \frac{b_2}{a_2}, \tag{6.94}$$

where $\omega_0 = V(\chi_1, T_{j-1}, S) z_1 e^{-z_1} < 0$ *is constant. Also this scheme preserves no-arbitrage.*

Proof. Obviously, at $K = K_1$ we have $\chi_1 = \log(K_1/S)$, $V(\chi, T_{j-1}, S) = V(\chi_1, T_{j-1}, S) \equiv V_1$, therefore, assuming the strike K_1 is out of the money

$$\omega_0 = V_1 z_1 e^{-z_1} < 0. \tag{6.95}$$

As this is described in detail in [Itkin and Lipton (2018)], the no-arbitrage conditions for the Put price read

$$P > 0, \quad P_K > 0, \quad P_{K,K} > 0.$$

Based on Eq.(6.82) and the definition of V in Eq.(6.45), the Put price at this interval can be represented as

$$P(K, T_{j-1}, S) = \omega_0 e^z / z = \omega_0 e^{b_2/a_2} \frac{K/S}{\log(K/S) + b_2/a_2}. \tag{6.96}$$

As at this interval $v = b_2 + a_2 \log(K/S)$, and it was assumed that $K < S$, we must have $a_2 < 0$. Accordingly, to have a positive Put price we require $\omega_0 < 0$. This is consistent with the value of ω_0 introduced in Eq.(6.95).

Differentiating Eq.(6.96) on K, and then again, we obtain

$$P_K' = \frac{\omega_0 a_2}{S} e^{b_2/a_2} \frac{b_2 - a_2 + a_2 \log(K/S)}{(b_2 + a_2 \log(K/S))^2} > 0, \tag{6.97}$$

$$P_K'' = -\omega_0 \frac{a_2^2}{KS} e^{b_2/a_2} \frac{b_2 - 2a_2 + a_2 \log(K/S)}{(b_2 + a_2 \log(K/S))^3} > 0.$$

Thus, the proposed scheme can be used for interpolation because it provides correct Put option prices at $K = K_1$ and $K \to 0$, and is monotone in K. Moreover, it preserves no-arbitrage. □

Now the integral in Eq.(6.51) can be computed in closed form

$$I_{12}(\chi) = y_2(\chi) \int \frac{y_1(\chi)c(\chi)}{(b_2 + a_2\chi)W}d\chi - y_1(\chi) \int \frac{y_2(\chi)c(\chi)}{(b_2 + a_2\chi)W}d\chi,$$
(6.98)

$$\int \frac{y_1(\chi)c(\chi)}{(b_2 + a_2\chi)W}d\chi = \xi_0 \int z^{-1}U(\alpha_1, \beta_1, z)dz = \xi_0 U_{-1}(0; \alpha_1, \beta_1, z),$$

$$\int \frac{y_2(\chi)c(\chi)}{(b_2 + a_2\chi)W}d\chi = \xi_0 \int e^z z^{-1}U(\beta_1 - \alpha_1, \beta_1, -z)dz$$

$$= -\xi_0 U_{-1}(-1; \beta_1 - \alpha_1, \beta_1, z),$$

$$\xi_0 = (-1)^{\beta_1 - \alpha_1}\omega_0 a_2^{-\beta_1}.$$

Representation of functions $U_{-1}(-1; \beta_1 - \alpha_1, \beta_1, z)$, $U_{-1}(0; \alpha_1, \beta_1, z)$ via the Meijer G-function is given in Eq.(6.92). Again, it can be easily verified that at $K \to 0$, and so $z \to -\infty$, the integral $I_{12}(\chi)$ vanishes.

6.4.4   *Special case $z \approx 1$ or $|v/b_2| \ll 1$*

This case occurs when at the interval $[K_i, K_{i+1}]$ for some $i \in [1, n_j]$ coefficients a_2, b_2 are such that either $|1 - z_i| \ll 1$ or $|1 - z_{i+1}| \ll 1$. Suppose, e.g. that $z_{i+1} = 1 + \epsilon$ with $0 < \epsilon \ll 1$. As shown in the next section, then we can introduce a ghost point K_* such that $z_* = 1 - \epsilon$. So at the interval $[K_*, K_{i+1}]$ we will use the numerically stable solution in Eq.(6.65), while at the interval $[K_i, K_*]$ — the regular solution in Eq.(6.63). Same construction could be provided if $z_i = 1 - \epsilon$.

At the interval $z \in [1 - \epsilon, 1 + \epsilon]$ where the values of z are close to singularity of the Hypergeometric function at $z = 1$ there are two ways to construct the solution. First, one can build an asymptotic solution using v/b_2 as a small parameter, because at $z \to 1$ we have $v/b_2 = (b_2 + a_2 x)/b_2 = 1 - z \to 0$. As shown in [Carr and Itkin (2018a)], this can be done, e.g., using the method of boundary functions, [Vasil'eva *et al.* (1995)].

Alternatively, it follows from Eq.(6.65) that $y_1(z) \to 1$, $y_2(z) \to 0$ at $z \to 1$. Therefore, these solutions have a regular behavior in the vicinity of $z = 1$. So all we need to do is to propose a suitable no-arbitrage interpolation to make computation of the source term in Eq.(6.58) tractable. This interpolation is constructed in 6.4.1.

Thus, based on Eq.(6.74) and Eq.(6.65) we need to compute 2 integrals

$$\mathcal{J}_1(x) = \int \frac{y_1(z)c(z)}{(1 - z)z^2 W(z)}dz, \qquad \mathcal{J}_2(x) = \int \frac{y_2(z)c(z)}{(1 - z)z^2 W(z)}dz, \quad (6.99)$$

$$y_1(z) = z^m {}_2F_1(m-1, m, 2m-c; 1-z), \quad c(z) = V(z, T_{j-1}, S),$$
$$y_2(z) = z^m (1-z)^{c-2m+1} {}_2F_1(c-m+1, c-m, c-2m+2; 1-z),$$

$$W(y_1(z), y_2(z)) = \omega_1 (1-z)^{c-2m} z^{2m-c}, \quad \omega_1 = -\frac{a_2(2m-1-c)}{b_2}.$$

The integral $\mathcal{J}_2(x)$ can be found in closed form, and the result reads

$$\mathcal{J}_2(x) = \bar{\gamma}_0 \mathcal{J}_{2,0}(x) + \gamma_1 \mathcal{J}_{2,1}(x) + \bar{\gamma}_2 \mathcal{J}_{2,1}(x), \qquad (6.100)$$

$$\mathcal{J}_{2,0}(x) = \frac{\pi}{\omega_1} \csc(\pi c) z^{-m} \Gamma(c-2m+2) \left[\frac{z^{c-1} {}_2F_1(c-m-1, c-m+1; c; z)}{(c-m-1)\Gamma(c)\Gamma(1-m)\Gamma(2-m)} \right.$$
$$\left. + \frac{{}_2F_1(2-m, -m; 2-c; z)}{m\Gamma(2-c)\Gamma(c-m)\Gamma(c-m+1)} \right],$$

$$\mathcal{J}_{2,1}(x) = \frac{\pi}{(m-1)\omega_1} \csc(\pi c) z^{-m} \Gamma(c-2m+2)$$
$$\cdot \left[\frac{z(c-m) {}_2F_1(1-m, 1-m; 2-c; z)}{\Gamma(2-c)\Gamma(c-m+1)^2} - \frac{z^c {}_2F_1(c-m, c-m; c; z)}{(c-m)\Gamma(c)\Gamma(1-m)^2} \right],$$

$$\mathcal{J}_{2,2}(x) = \frac{\Gamma(c-2m+2)}{\omega_1 \Gamma(1-m)\Gamma(2-m)\Gamma(c-m)\Gamma(c-m+1)}$$
$$\times G_{3,3}^{2,3} \left(\begin{matrix} 1,1,2 \\ 2-m, \ c-m+1, \ 0 \end{matrix} \middle| z \right).$$

The integral $\mathcal{J}_1(x)$ with the use of no-arbitrage interpolation defined in Eq.(6.78) reads

$$\mathcal{J}_1(x) = \omega_1^{-1} \int (1-z)^{-c+2m-1} z^{c-m-2} {}_2F_1(m-1, m; 2m-c; 1-z)$$
$$\cdot (\bar{\gamma}_0 + \gamma_1 z + \bar{\gamma}_2 z^2) dz.$$

This integral can be computed as follows. We remind that $z \in [1-\epsilon, 1+\epsilon]$, $|\epsilon| \ll 1$. Therefore, the term z^k, $k \in \mathbb{R}$ can be expanded into series around $z = 1$ to obtain

$$z^k = \sum_{i=0}^{\infty} (-1)^i \binom{k}{i} (1-z)^i.$$

Then $\mathcal{J}_1(x)$ takes the form

$$\mathcal{J}_1(x) = \omega_1^{-1} \left\{ \bar{\gamma}_0 \sum_{i=0}^{\infty} (-1)^i \binom{c-m-2}{i} \right.$$
$$\cdot \int (1-z)^{i-c+2m-1} {}_2F_1(m-1, m; 2m-c; 1-z) dz \qquad (6.101)$$
$$+ \gamma_1 \sum_{i=0}^{\infty} (-1)^i \binom{c-m-1}{i}$$

$$\cdot \int (1-z)^{i-c+2m-1} \, _2F_1(m-1,m;2m-c;1-z)dz$$

$$+\bar{\gamma}_2 \sum_{i=0}^{\infty} (-1)^i \binom{c-m}{i} \int (1-z)^{i-c+2m-1}$$

$$\times \, _2F_1(m-1,m;2m-c;1-z)dz \bigg\}$$

$$= \omega_1^{-1} \sum_{i=0}^{\infty} \nu_i \int (1-z)^{i-c+2m-1} \, _2F_1(m-1,m;2m-c;1-z)dz,$$

$$= \omega_1^{-1} \sum_{i=0}^{\infty} \frac{\nu_i}{c-i-2m}(1-z)^{-c+i+2m}$$

$$\times \, _3F_2 \begin{bmatrix} m-1, \ m, \ 2m-c+i \\ 2m-c, \ 2m+i-c+1 \end{bmatrix}; 1-z \bigg],$$

$$\nu_i = (-1)^i \left[\bar{\gamma}_0 \binom{c-m-2}{i} + \gamma_1 \binom{c-m-1}{i} + \bar{\gamma}_2 \binom{c-m}{i} \right].$$

The exponent $-c+i+2m = i+b_1/b_2$ is always positive if $b_2 > 0$ in the vicinity of $z = 1$. According to 6.4.1, this condition on b_2 is valid if $1-\epsilon \le z < 1$. Therefore, 2–3 terms in the expansion Eq.(6.101) provide the sufficient accuracy in computation of the integral. However, this is also true when $1+\epsilon > z > 1$ (and so b_2 is negative) which implies that the entire exponent is also negative, at least at low i. This is because the behavior of the product $(1-z)^{i-c+2m} \, _3F_2 \begin{bmatrix} m-1, \ m, \ 2m-c+i \\ 2m-c, \ 2m+i-c+1 \end{bmatrix}; 1-z \bigg]$ is regular even in this case.

In a similar manner the source terms for other models of the local variance/volatility considered in previous sections could be computed in closed form. We leave this exercise to the reader.

6.5 Smile calibration for a single term

Calibration problem for the local volatility model is described in [Carr and Itkin (2018a)] as well as the construction of the solution for the entire smile. Here we follow the same approach, and, therefore, provide just some short comments specific to the GLVG model. Again, as an example consider the case where the local variance is a piecewise linear function of strike. Calibration for the other cases considered in Section 6.3 can be done in a similar manner.

A general calibration problem we need to solve is: given market quotes of Call and/or Put options corresponding to various strikes $\{K\} := K_j$, $j \in [1, N]$ and same maturity T_i, find the local variance function $v(x)$ such that these quotes solve equations in Eq.(6.37), Eq.(6.43).

Suppose that the Put prices for $T = T_j$ are known for n_j ordered strikes. The location of those strikes on the x line is schematically depicted in Fig. 6.1.

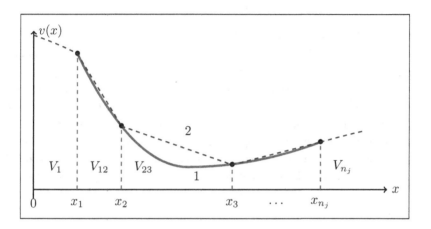

Figure 6.1: Schematic construction of the combined solution in $x \in \mathbb{R}^+$: 1 (solid line) — the real (unknown) local variance curve, 2 (dashed line) — a piecewise linear solution. At $x > x_{n_j}$ and $x < x_1$ the dashed line is $b_2 + a_2 \log(x)$.

Recall that the general form of the solution is given in Eq.(6.51) which at every interval $x_{i-1} \leq x \leq x_i$ and $T = T_j$ can be represented as

$$V(x) = C_{j,i}^{(1)} y_1(x) + C_{j,i}^{(2)} y_2(x) + I_{12}(x). \tag{6.102}$$

Here for better readability we changed the notation of two integration constants which belong to the i-th interval in x and j-th maturity to $C_{j,i}^{(1)}, C_{j,i}^{(2)}$.

Similar to [Carr and Itkin (2018a)], we assume continuity of the options price and its first derivative at every node $i = 1, \ldots, n_j$. We also supplement this by two additional conditions: the first one is given by Eq.(6.49), and the other one is that at every node the solution $P(S, T_j, K_i)$ must coincide with a given market quote for the pair (T_j, K_i). So together this provides four equations for four unknown variables $v_{j,i}^0, v_{j,i}^1, C_{j,i}^{(1)}, C_{j,i}^{(2)}$:

$$P_i(x)|_{x=x_i} = P_{i+1}(x)|_{x=x_i}, \tag{6.103}$$

$$P_i(x)|_{x=x_i} = P_{market}(x_i),$$

$$\frac{\partial P_{i+1}(x)}{\partial x}\bigg|_{x=x_i} = \frac{\partial P_i(x)}{\partial x}\bigg|_{x=x_i},$$

$$v_{j,i}^0 + v_{j,i}^1 x_i = v_{j,i+1}^0 + v_{j,i+1}^1 x_i, \quad i = 1, \ldots, n_j.$$

The Eq.(6.103) is a system of $4n_j$ nonlinear equations with respect to $4(n_j + 1)$ variables $v_{j,i}^0, v_{j,i}^1, C_{j,i}^{(1)}, C_{j,i}^{(2)}$. Therefore we need 4 additional conditions to unquietly solve it.

To this end observe that the constants $C_{j,1}^{(2)}, C_{j,n_j}^{(2)}$ could be determined based on the boundary conditions in Eq.(6.47). Indeed, at $K \to 0$ function $y_2(\chi)$ in Eq.(6.54) vanishes (as $a_2 < 0$ at this interval), but not $y_1(x)$. Therefore, to obey the vanishing boundary condition in Eq.(6.47) we must set $C_{j,1}^{(1)} = 0$. As that was already discussed, the source term in Eq.(6.98) also vanishes in this limit. Therefore, the solution in Eq.(6.54) with the source term in Eq.(6.98) and $C_{j,1}^{(2)} = 0$ obeys the boundary condition at $z \to 0$.

At $K \to \infty$ based on representation of the solution in Eq.(6.54) with $a_2 > 0$ at this interval, similarly we must set $C_{j,n_j}^{(2)} = 0$, as the solution $y_2(x)$ in Eq.(6.54) diverges at $z \to \infty$.

The remaining two additional conditions could be set in many different ways. Here we rely on traders intuition about the asymptotic behavior of the volatility surface at strikes close to zero and infinity. According to our construction, they are determined by $v_{j,0}^1$ and v_{j,n_j}^1. Therefore, we assume these coefficients to be somehow known, i.e. consider them as the given parameters of our model.

Overall, by solving the nonlinear system of equations Eq.(6.103) we find the final solution of our problem. This can be done by using standard methods, and, thus, no optimization procedure is necessary. However, a good initial guess still would be helpful for a better (and faster) convergence. Construction of such a guess is described in [Carr and Itkin (2018a)]. Also note that this system has a block-diagonal structure where each block is a 2x2 matrix. Therefore, it can be easily solved with the linear complexity $O(n_j)$.

When computing the first derivatives, we take into account that derivatives of Hypergeometric functions belong to the same class of functions, since, [Abramowitz and Stegun (1964)]

$$\frac{\partial}{\partial z} {}_2F_1\left(a,b,c,z\right) = \frac{ab}{c} {}_2F_1\left(a+1,b+1,c+1,z\right),$$

$$\frac{\partial}{\partial z} {}_3F_2\begin{bmatrix} a, & b, & c \\ & d, & e \end{bmatrix}; z = \frac{abe}{cd} {}_3F_2\begin{bmatrix} a+1, & b+1, & c+1 \\ & d+1, & e+1 \end{bmatrix}; z.$$

Same is true for the Meijer G-function. For instance,

$$\frac{\partial}{\partial z} G_{2,3}^{2,2}\left(\begin{matrix} \nu, & 1+\alpha_1-\beta_1 \\ 0, & 1-\beta_1, & \nu-1 \end{matrix}\middle| -z\right) = \frac{\Gamma(1-\alpha_1)\Gamma(\beta_1-\alpha_1)}{z} U(\beta_1-\alpha_1,\beta_1,-z)$$

(6.104)

$$+ (\nu-1)G_{2,3}^{2,2}\left(\begin{matrix} \nu, & 1+\alpha_1-\beta_1 \\ 0, & 1-\beta_1, & \nu-1 \end{matrix}\middle| -z\right).$$

Therefore, computing derivatives of the solution does not cause any new technical problem.

6.5.1 *Special case* $|1 - z_i| \ll 1$ *at some node* K_i, $i \in [1, n_j]$

Without loss of generality suppose that $z_i = 1 - \epsilon$ and $z_{i+1} \gg 1 + \epsilon$ with $0 < \epsilon \ll 1$. The other case $z_i = 1 + \epsilon$ and $z_{i-1} \ll 1 - \epsilon$ could be treated in a similar way. Then let us introduce a ghost point K_* such that $z_* = 1 + \epsilon$. So at the interval $[K_i, K_*]$ we will use the numerically stable solution in Eq.(6.65), while at the interval $[K_*, K_{i+1}]$ — the regular solution in Eq.(6.63).

Since K_* is the ghost point, we don't have a market quote available at K_*. All we can say is that yet we assume the local variance/volatility to be a piecewise linear function of K at $[K_*, K_{i+1}]$ and $[K_i, K_*]$. It has to be continuous but with a possible jump in skew at K_*.

Since a market quote at K_* is not available, we can replace it with any reasonable value. For instance, an interpolated value between market quotes at K_i, K_{i+1} could be used obtained by using no-arbitrage interpolation.[5] Then we obtain four equations for $C_{j,*}^{(1)}, C_{j,,*}^{(2)}, v_{j,*}^0, v_{j,*}^1$

$$P_i(x)|_{x=x_i} = P_*(x)|_{x=x_i}, \qquad (6.105)$$

$$P_*(x)|_{x=x_*} = P_{interp}(x)|_{x=x_*},$$

$$\frac{\partial P_*(x)}{\partial x}\bigg|_{x=x_i} = \frac{\partial P_i(x)}{\partial x}\bigg|_{x=x_i},$$

$$v_{j,i}^0 + v_{j,i}^1 x_* = v_{j,*}^0 + v_{j,*}^1 x_*, \qquad i = 1, \ldots, n_j.$$

that should be added to Eq.(6.103). Solving this new combined linear system in the same way as we did it for Eq.(6.103) we find the values of all unknown $C_{j,i}^{(1)}, C_{j,,i}^{(2)}, v_{j,i}^0, v_{j,i}^1$ where now $i \in \{[1, n_j] \cup *\}$.

---

[5]Despite it looks attractive, we cannot require $v_{j,i}^1 = v_{j,*}^1$ since this also gives rise to $v_{j,i}^0 = v_{j,*}^0$. However, $v_{j,i}^0$ changes sign at $z = 1$.

6.6   Discussion

First, let us mention that in many practical calculations either coefficients $a_2 = v_{j,i}^1$ at some i, or both $b_2 = v_{j,i}^0$, $a_2 = v_{j,i}^1$ (see, for instance, Eq.(6.57)) are small. Of course, in that case the general solution Eq.(6.63) remains valid. However, when computing the values of Hypergeometric functions numerically, the errors significantly grow in such a case. This is especially pronounced when computing the source term integral I_{12}. The main point is that either the Hypergeometric function takes a very small value, and then the constants $C_{j,i}^{(1)}, C_{j,i}^{(2)}$ should be very large to compensate, or vice versa. Resolution of this issue requires high-precision arithmetics, and, which is more important, taking into account many terms in a series representation of the Hypergeometric functions, which significantly slows down the total performance of the method.

To eliminate these problems we can look at asymptotic solutions of Eq.(6.57) taking into account the existence of small parameters from the very beginning. This approach was successfully elaborated on in [Itkin and Lipton (2018); Carr and Itkin (2018a)], so we don't describe it here in detail.

In [Carr and Itkin (2018a)] we calibrated the ELVG model, e.g. to the data set taken from [Balaraman (2016)]. In that paper an implied volatility surface of S&P500 is presented, and the local volatility surface is constructed using the Dupire formula. We took data for the first 12 maturities and all strikes as they are given in [Balaraman (2016)]. Our results demonstrated high accuracy and speed of calibration.

When doing so, a technical note should be made. We mentioned already that in our model for every term the slopes of the smile at strikes close to zero, $v_{j,0}^1$ and infinity, v_{j,n_j}^1 are free parameters of the model. So often traders have an intuition about these values. However, in our numerical experiments we setup them just using some plausible test values. In particular, in [Carr and Itkin (2018a)] for the sake of simplicity for all smiles we used $v_{j,0}^1 = -0.3$, and $v_{j,n_j}^1 = 0.1$. Accordingly, for the instantaneous variance $v_j(x_i) = p_j(v_{j,i}^0 + v_{j,i}^1 \log(x_i))/2$ the slopes at both zero and plus infinity are time-dependent and can be computed by using this definition.

As a numerical solver for the system of linear equations we used the standard Matlab *fsolve* function, and utilized a "trust-region-dogleg" algorithm. Parameter "TypicalX" has to be chosen carefully to speedup calculations.

In [Carr and Itkin (2018b)] we repeated this test, but now using the GLVG instead of the ELVG. The results look same as in Fig.5 of [Carr and Itkin (2018a)], i.e. the quality of the fit is same, and performance of the method is almost same. But the conclusion of [Carr and Itkin (2018a)] remains intact, namely that performance of this model is much better than that reported in both [Itkin (2015)] and [Itkin and Lipton (2018)].

Therefore, a natural question would be: which flavor of the Local Variance Gamma model — arithmetic or geometric one is preferable. Perhaps, if the ultimate goal is fast calibration of the given smile, both could be used interchangeably, and both are capable to provide a good and fast fit. However, for modeling option prices the difference between the geometric and arithmetic LVG models is of the same kind as between the Bachelier and Black-Scholes models. So, for instance, for modeling stock prices the latter would be preferable, while for modeling interest rates the former could provide negative values, which nowadays is a desirable feature.

Bibliography

Abramowitz, M. and Stegun, I. (1964). *Handbook of Mathematical Functions* (Dover Publications, Inc.).

Ahoniemi, K. (2009). *Modeling and Forecasting Implied Volatility*, Ph.D. thesis, Helsinki School of Economics.

Alentorn, A. (2004). Modelling the implied volatility surface: An empirical study for FTSE options, Tech. rep., Centre of Computational Finance and Economics Agents, University of Essex.

Alexander, C. (2001). Principles of the skew, *Risk* **14**, 1, pp. 529–532.

Andreasen, J. and Huge, B. (2011). Volatility interpolation, *Risk Magazine*, pp. 76–79.

Andreou, P., Charalambous, C., and Martzoukos, S. (2014). Assessing the performance of symmetric and asymmetric implied volatility functions, *Review of Quantitative Finance and Accounting* **42**, 3, pp. 373–397.

Antonov, A., Konikov, M., and Spector, M. (2019). A new arbitrage-free parametric volatility surface, https://papers.ssrn.com/sol3/papers.cfm?abstract_id=3403708&download=yes, SSRN:3403708.

Askey, R. and Daalhuis, A. B. O. (2010). Generalized hypergeometric function, in F. Olver, D. Lozier, R. Boisvert, and C. Clark (eds.), *NIST Handbook of Mathematical Functions* (Cambridge University Press).

Balaraman, G. (2016). Modeling Volatility Smile and Heston Model Calibration Using QuantLib Python, Available at http://gouthamanbalaraman.com/blog/volatility-smile-heston-model-calibration-quantlib-python.html.

Bateman, H. and Erdélyi, A. (1953). *Higher Transcendental Functions, Bateman Manuscript Project California Institute of Technology*, Vol. 1 (McGraw-Hill).

Bergomi, L. (2016). *Stochastic Volatility Modeling*, CRC Financial Mathematics Series (Chapman and Hall).

Biscamp, L. (2008). Private communication.

Bochner, S. (1949). Diffusion equation and stochastic processes, in *Proceedings of the National Academy of Sciences, USA*, Vol. 35, pp. 368–370.

Borovkova, S. and Parmana, F. (2009). Implied volatility in oil markets, *Computational Statistics and Data Analysis* **53**, 6, pp. 2022–2039.

Bossu, S. (2014). *Advanced Equity Derivatives: Volatility and Correlation*, Wiley Finance (Wiley), ISBN 9781118774717.

Breeden, D. and Litzenberger, R. (1978). Prices of state-contingent claims implicit in option prices, *Journal of Business* **51**, 4, pp. 621–651.

Brigo, D. and Mercurio, F. (2006). *Interest Rate Models — Theory and Practice with Smile, Inflation and Credit*, 2nd edn. (Springer Verlag).

Carr, P. (2004). Implied vol constraints, Available at http://www.javaquant.net/papers/impvolconstrs3.pdf.

Carr, P. (2014a). Options as optimizations: A dual approach to derivatives pricing, Quant USA (New York).

Carr, P. (2014b). Private communication.

Carr, P., Fisher, T., and Ruf, J. (2013). Why are quadratic normal volatility models analytically tractable? *SIAM Journal on Financial Mathematics* **4**, 1, pp. 185–202.

Carr, P. and Itkin, A. (2018a). An expanded local variance gamma model, Available at https://arxiv.org/pdf/1802.09611.pdf.

Carr, P. and Itkin, A. (2018b). Geometric local variance gamma model, Available at https://arxiv.org/abs/1809.07727.

Carr, P. and Madan, D. (2005). A note on sufficient conditions for no arbitrage, *Finance Research Letters* **2**, pp. 125–130.

Carr, P. and Nadtochiy, S. (2014). Local variance gamma and explicit calibration to option prices, Available at https://arxiv.org/abs/1308.2326.

Carr, P. and Nadtochiy, S. (2017). Local Variance Gamma and explicit calibration to option prices, *Mathematical Finance* **27**, 1, pp. 151–193.

Carr, P. and Pelts, G. (2015). Duality, deltas, and derivatives pricing, in *Steve Shreveś 65th Birthday Conference*, http://www.math.cmu.edu/CCF/CCFevents/shreve/abstracts/P.Carr.pdf.

Carr, P. and Wu, L. (2010). A new framework for analyzing volatility risk and premium across option strikes and expiries, http://papers.ssrn.com/sol3/papers.cfm?abstract_id=1701685.

Castagna, A. (2010). *FX Options and Smile Risk* (Wiley).

Ciliberti, S., Bouchaud, J., and Potters, M. (2008). Smile dynamics — a theory of the implied leverage effect, (arXiv, 0809.3375v1 [q-fin.PR]).

Coleman, T., Kim, Y., Li, Y., and Verma, A. (2001). Dynamic hedging with a deterministic local volatility function model, *The Journal of Risk* **4**, 1, pp. 63–89.

Cont, R. and Fonseca, J. (2002). Dynamics of implied volatility surfaces, *Quantitative Finance*, 2, pp. 45–60.

Corcuera, J., Guillaume, F., Leoni, P., and Schoutens, W. (2009). Implied Lévy volatility, *Quantitative Finance* **9**, 4, pp. 383–393.

Cox, J. and Rubinstein, M. (1985). *Options Markets* (Prentice-Hall).

Daglish, T., Hull, J., and Suo, W. (2007). Volatility surfaces: theory, rules of thumb, and empirical evidence, *Quantitative Finance* **7**, 5, pp. 507–524.

De Marco, S., Friz, P., and Gerhold, S. (2013). Rational shapes of local volatility, *Risk*, 2, pp. 82–87.

Denneberg, D. (1990). Distorted probabilities and insurance premiums. *Methods of Operations Research* **63**, 2, pp. 21–42.

Derman, E. (1999). Regimes of volatility some observations on the variation of S&P 500 implied volatilities, Quantitative strategies research notes, Goldman Sacks.

Derman, E. and Kani, I. (1994a). Riding on a smile, *RISK*, pp. 32–39.

Derman, E. and Kani, I. (1994b). The volatility smile and its implied tree, Tech. rep., GS Quantitative Strategies Research Notes.

Derman, E. and Kani, I. (1998). Stochastic implied trees: Arbitrage pricing with stochastic term and strike structure of volatility, *International Journal of Theoretical and Applied Finance* **1**, 1, pp. 61–110.

Derman, E., Miller, M., and Park, D. (2016). *The Volatility Smile*, Wiley Finance (Wiley), ISBN 9781118959169.

Dumas, B., Fleming, J., and Whaley, R. (1998). Implied volatility functions: Empirical tests, *The Journal of Finance* **LIII**, 6, pp. 2059–2106.

Dupire, B. (1994). Pricing with a smile, *Risk* **7**, pp. 18–20.

Ekström, E. and Tysk, J. (2012). Dupire's equation for bubles, *International Journal of Theoretical and Applied Finance* **15**, 6, pp. 1250041–1250053.

Falck, M. and Deryabin, M. (2017). Local variance gamma revisited, Available at https://papers.ssrn.com/sol3/papers.cfm?abstract_id=2659728.

Fengler, M. (2005). Arbitrage-free smoothing of the implied volatility surface, Tech. Rep. SFB 649 Discussion Paper, Trading & Derivatives.

Fengler, M., Härdle, W., and C.Villa (2003). A common principal component approach, *Review of Derivatives Research* **6**, 3, pp. 179–202.

Gasull, A. and Utzet, F. (2014). Approximating mills ratio, *Journal of Mathematical Analysis and Applications* **420**, 2, pp. 1832–1863.

Gatheral, J. (1999). The volatility skew: Arbitrage constraints and asymptotic behaviour, Tech. rep., Merrill Lynch.

Gatheral, J. (2002). Stochastic volatility and local volatility, http://web.math.ku.dk/~rolf/teaching/ctff03/Gatheral.1.pdf.

Gatheral, J. (2004). A parsimonious arbitrage-free implied volatility parameterization with application to the valuation of volatility derivatives, Global Derivatives And Risk Management.

Gatheral, J. (2006). *The volatility surface* (Wiley finance).

Gatheral, J., Hsu, E., Laurence, P., Ouyang, C., and Wang, T. (2012). Asymptotics of implied volatility in local volatility models, *Mathematical Finance* **22**, 4, pp. 591–620.

Gatheral, J. and Jacquier, A. (2011). Convergence of Heston to SVI, *Quantitative Financet* **11**, 8, pp. 1129–1132.

Gatheral, J. and Jacquier, A. (2014). Arbitrage-free SVI volatility surfaces, *Quantitative Finance* **14**, 1, pp. 59–71.

Gerhold, S. and Friz, P. (2015). Extrapolation analytics for Dupire's local volatility, in *Large Deviations and Asymptotic Methods in Finance, Springer Proceedings in Mathematics & Statistics*, Vol. 110 (Springer), pp. 273–286.

Hagan, P., Kumar, D., A, A. L., and Woodward, D. (2002). Managing smile risk, *Wilmott magazine*, pp. 84–108.

Hansen, N. (2008). CMA-ES source code, https://www.lri.fr/~hansen/cmaes_ inmatlab.html, available at https://www.lri.fr/~hansen/cmaes_inmatlab. html.

Heston, S. (1993). A closed-form solution for options with stochastic volatility with applications to bond and currency options, *Review of Financial Studies* **6**, pp. 327–343.

Hodges, H. (1996). Arbitrage bounds of the implied volatility strike and term structures of European-Style Options, *Journal of Derivatives* **3**, 4, pp. 23–32.

Homescu, C. (2011). Implied volatility surface: Construction methodologies and characteristics, SSRN, 1882567.

Homescu, C. (2014). Local stochastic volatility models: Calibration and pricing, SSRN, 2448098.

Hull, J. and White, A. (2015). A generalized procedure for building trees for the short rate and its application to determining market implied volatility functions, *Quantitative Finance* **15**, 3, pp. 443–454.

Hull, J. C. (1997). *Options, Futures, and other Derivative Securities*, 3rd edn. (Prentice-Hall, Inc., Upper Saddle River, NJ).

Itkin, A. (2015). To sigmoid-based functional description of the volatility smile, *North American Journal of Economics and Finance* **31**, pp. 264–291.

Itkin, A. (2017). *Pricing derivatives under Lévy models*, 1st edn., no. 12 in Pseudo-Differential Operators (Birkhauser, Basel).

Itkin, A. (2018). Nonlinear PDEs risen when solving some optimization problems in finance, and their solutions, *Journal of Computational Finance* **21**, 4, pp. 1–21.

Itkin, A. (2019). Deep learning calibration of option pricing models: some pitfalls and solutions, https://arxiv.org/abs/1906.03507, arxiv: 1906.03507.

Itkin, A. and Lipton, A. (2018). Filling the gaps smoothly, *Journal of Computational Sciences* **24**, pp. 195–208.

Kienitz, J. and Caspers, P. (2017). *Interest Rate Derivatives Explained: Term Structure and Volatility Modelling, Financial Engineering Explained*, Vol. 2, 1st edn. (Palgrave Macmillan UK).

Kienitz, J. and Wetterau, D. (2012). *Financial Modelling: Theory, Implementation and Practice with MATLAB Source*, The Wiley Finance Series (Wiley), ISBN 9781118413319.

Kotzé, A., Labuschagne, C., Nair, M., and Padayachi, N. (2013). Arbitrage-free implied volatility surfaces for options on single stock futures, *North American Journal of Economics and Finance* **26**, pp. 380–399.

Kuznetsov, A. (2013). On the convergence of the Gaver-Stehfest algorithm, *SIAM J. Numerical Analysis* **51**, 6, pp. 2984–2998.

le Roux, M. (2007). A long-term model of the dynamics of the S&P500 implied volatility surface, *North American Actuarial Journal* **11**, 4, pp. 61–75.

Ledoit, O., Santa-Clara, P., and Yan, S. (2002). Relative pricing of options with stochastic volatility, Tech. rep., Eller College of Business and Public Administration, The University of Arizona.

Lee, R. (2004). The moment formula for implied volatility at extreme strikes, *Mathematical Finance*. **14**, 3, pp. 469–480.

Levy, H. (1985). Upper and lower bounds of put and call option value: Stochastic dominance approach, *The Journal of Finance* **40**, 4, pp. 1197–1217.

Lindholm, L. (2014). Calibration of local volatility surfaces under PDE constraints, https://www.diva-portal.org/smash/get/diva2:764597/FULLTEXT01.pdf.

Lipton, A. (2001). *Mathematical Methods For Foreign Exchange: A Financial Engineer's Approach* (World Scientific).

Lipton, A. (2002). The vol smile problem, *Risk*, pp. 61–65.

Lipton, A. and Sepp, A. (2011a). Credit value adjustment in the extended structural default model, in *The Oxford Handbook of Credit Derivatives* (Oxford University), pp. 406–463.

Lipton, A. and Sepp, A. (2011b). Filling the gaps, *Risk Magazine*, pp. 86–91.

Lörinczi, J., Hiroshima, F., and Betz, V. (2011). *Feynman-Kac-Type Theorems and Gibbs Measures on Path Space*, no. 34 in De Gruyter Studies in Mathematics (Walter de Gruyter GmbH & Co, Berlin/Boston).

Madan, D., Carr, P., and Chang, E. (1998). The variance gamma process and option pricing, *European Finance Review* **2**, pp. 79–105.

Madan, D. and Seneta, E. (1990). The variance gamma (V.G.) model for share market returns, *Journal of Business* **63**, 4, pp. 511–524.

Medvedev, A. and Scaillet, O. (2008). Pricing american options under stochastic volatility and stochastic interest rates, Tech. Rep. 429, National Centre of Competence in Research, Financial Valuation and Risk Management.

Natenberg, S. (1994). *Option Volatility & Pricing: Advanced Trading Strategies and Techniques* (McGraw-Hill Education), ISBN 9781557384867.

Nayfeh, A. H. (2000). *Perturbation methods* (John Wiley & Sons).

Ng, E. and Geller, M. (1970). On some indefinite integrals of confluent hypergeometric functions, *Journal of research of the Notional Bureau of Standards — B. Mathematical Sciences* **748**, 2, pp. 85–98.

Nuclear Phynance (2007). Available at http://www.nuclearphynance.com/Show%20Post.aspx?PostIDKey=98444.

Olver, F. (1997). *Asymptotics and Special Functions* (AKP Classics).

Polyanin, A. and Zaitsev, V. (2003). *Handbook of exact solutions for ordinary differential equations*, 2nd edn. (CRC Press Company, Boca Raton, London, New York, Washington, D.C.).

Prudnikov, A. P., Brychkov, Y. A., and Marikov, O. I. (1986). *Integrals and Series* (Gordon and Breach).

Rebonato, R. (2004). *Volatility and Correlation: The Perfect Hedger and the Fox* (Wiley).

Revuz, D. and Yor, M. (1999). *Continuous Martingales and Brownian Motion*, 3rd edn. (Springer, Berlin, Germany).

Risken, H. and Haken., H. (1989). *The Fokker-Planck Equation: Methods of Solution and Applications Second Edition* (Springer).

Romo, J. (2011). Fitting the skew with an analytical local volatility function, *International Review of Applied Financial Issues and Economics* **3**, pp. 721–736.

Romo, J. (2014). Dynamics of the implied volatility surface. theory and empirical evidence, *Quantitative Finance* **14**, 10, pp. 1829–1837.

Rosenberg, J. (2000). Implied volatility functions: A reprise, *Journal of Derivatives* **7**, pp. 51–64.

Rouah, F. D. (2001). Derivation of local volatility, https://frouah.com/finance%20notes/Dupire%20Local%20Volatility.pdf.

Schweizer, M. and Wissel, J. (2008). Arbitrage-free market models for option prices: The multi-strike case, *Finance and Stochastics* **12**, 4, pp. 469–505.

Sehgal, S. and Vijayakumar, N. (2008). Determinants of implied volatility function on the nifty index options market: Evidence from India, *Asian Academy of Management Journal of Accounting and Finance* **4**, pp. 45–69.

Sepp, A. (2014). Empirical calibration and minimum-variance Delta under Log-Normal stochastic volatility dynamics, SSRN, 2387845.

Shreve, S. (1992). Martingales and the theory of capital-asset pricing, *Lecture Notes in Control and Information SCIENCES* **180**, pp. 809–823.

Simon, D. (2013). *Evolutionary Optimization Algorithms* (Wiley).

Sinclair, E. (2013). *Volatility Trading* (Wiley).

Soize, C. (1994). *The Fokker-Planck Equation for Stochastic Dynamical Systems and Its Explicit Steady State Solutions*, Advanced Series on Fluid Mechanics (World Scientific), ISBN 9789810217556.

Tompkins, R. (2001). Implied volatility surfaces: Uncovering regularities for options on financial futures, *The European Journal of Finance* **7**, 3, pp. 198–230.

Van Kampen, N. (2007). *Stochastic processes in physics and chemistry* (North Holland).

Vasil'eva, A., Butuzov, V., and Kalachov, L. (1995). *The boundary function method for singular perturbation problems*, Studies in Applied Mathematics (SIAM, Philadelphia).

von Seggern, D. (2007). *CRC Standard Curves and Surfaces with Mathematics*, 2nd edn. (CRC Press).

Wasow, W. (1987). *Asymptotic Expansions for Ordinary Differential Equations* (Dover Pubns).

Zhao, B. and Hodges, S. D. (2013). Parametric modeling of implied smile functions: a generalized SVI model, *Review of Derivative Research* **16**, 53–77.

Zheng, Y., Yang, Y., and Chen, B. (2019). Gated deep neural networks for implied volatility surfaces, https://arxiv.org/pdf/1904.12834.pdf, arxiv: 1904.12834.

Index

Printed in the United States
By Bookmasters